To my wife

METAL
CARBONYL
SPECTRA

ORGANOMETALLIC CHEMISTRY

A Series of Monographs

EDITORS

P. M. MAITLIS
THE UNIVERSITY
SHEFFIELD, ENGLAND

F. G. A. STONE
UNIVERSITY OF BRISTOL
BRISTOL, ENGLAND

ROBERT WEST
UNIVERSITY OF WISCONSIN
MADISON, WISCONSIN

BRIAN G. RAMSEY: Electronic Transitions in Organometalloids, 1969.

R. C. POLLER: The Chemistry of Organotin Compounds, 1970.

RUSSELL N. GRIMES: Carboranes, 1970.

PETER M. MAITLIS: The Organic Chemistry of Palladium, Volume I, Volume II—1971.

DONALD S. MATTESON: Organometallic Reaction Mechanisms of the Nontransition Elements, 1974.

RICHARD F. HECK: Organotransition Metal Chemistry: A Mechanistic Approach, 1974.

P. W. JOLLY AND G. WILKE: The Organic Chemistry of Nickel, Volume I, Organonickel Complexes, 1974. Volume II, Organic Synthesis, in preparation.

P. C. WAILES, R. S. P. COUTTS, AND H. WEIGOLD: Organometallic Chemistry of Titanium, Zirconium, and Hafnium, 1974.

U. BELLUCO: Organometallic and Coordination Chemistry of Platinum, 1974.

L. MALATESTA AND S. CENINI: Zerovalent Compounds of Metals, 1974.

P. S. BRATERMAN: Metal Carbonyl Spectra, 1975.

In preparation

R. P. A. SNEEDEN: Organochromium Compound.

THOMAS ONAK: Organoborane Chemistry.

METAL CARBONYL SPECTRA

P. S. BRATERMAN

Department of Chemistry,
University of Glasgow,
Scotland.

1975

ACADEMIC PRESS

LONDON   NEW YORK   SAN FRANCISCO

A Subsidiary of Harcourt Brace Jovanovich, Publishers

ACADEMIC PRESS INC. (LONDON) LTD.
24/28 Oval Road,
London NW1

United States Edition published by
ACADEMIC PRESS INC.
111 Fifth Avenue
New York, New York 10003

Copyright © 1975 by
ACADEMIC PRESS (INC.) LONDON LTD.

Library of Congress Catalog Card Number: 73–19001
ISBN: 0–12–125850–5

Printed in Great Britain by
ROYSTAN PRINTERS LIMITED
Spencer Court, 7 Chalcot Road
London NW1

Preface

This book is addressed to organometallic chemists who, like the author, do not think of themselves primarily as spectroscopists, but who nonetheless realise the need to make the best possible use of spectroscopic data. The mathematical development is descriptive rather than rigorous, and the use of matrix notation is restricted to Chapter 2 and short sections of Chapter 3. Some acquaintance with Group Theory is however assumed.

The greater part of the book is devoted to vibrational spectroscopy, reflecting the central position that this occupies both in routine characterisation and as a source of information about structural features and bonding. Although it would be neither possible nor desirable to refer to all reports of vibrational spectra for metal carbonyls, an attempt has been made to present a range of data extensive enough to act as a basis for comparison with new results.

The final section discusses the use of ultraviolet and photoelectron spectroscopy as well as carbon-13 n.m.r. methods for metal carbonyls. Although the information available is as yet scant, it is not too early to draw attention to the potential usefulness of these methods in metal carbonyl chemistry.

I wish to thank numerous teachers and colleagues, and in particular the research students who have worked with me in this area; Mrs. Frieda Lawrie and her associates for the collection of spectra and for generous help with the literature survey; and finally my wife, Michele, who has not only read and typed the manuscript, and assisted in preparing it for the press, but has made numerous helpful suggestions regarding presentation and style.

Note on Units

With the exception of the reciprocal centimetre as a measure of wavenumber, SI units are used throughout this work. Force constants are measured in Newtons metre^{-1}, and internuclear distances in pm. Some relationships between SI and other units are

$$1\,cm^{-1} = 10^{2c}\,Hz$$

(where c is the speed of light in m. sec^{-1})

$$= 2 \cdot 9979 \times 10^{10}\,Hz$$

$$1\,cm^{-1} = 1 \cdot 9864 \times 10^{-23}\,J, \text{ corresponding to}$$
$$11 \cdot 962\,J\,.\,mol^{-1}$$

$$= 1 \cdot 2398 \times 10^{-4}\,eV, \text{ corresponding to}$$
$$7 \cdot 4663 \times 10^{19}\,eV.\,mol^{-1}$$

Thus

$$2{,}000\,cm^{-1} = 5 \cdot 9958 \times 10^{13}\,Hz = 3 \cdot 9728$$
$$\times 10^{-20}J = 2 \cdot 4796 \times 10^{-1}\,eV,$$
$$\text{corresponding to } 23 \cdot 924\,kJ\,.\,mol^{-1}$$
$$\text{or } 1 \cdot 4933 \times 10^{23}\,eV.\,mol^{-1}$$

$$1\,md/\text{Å} = 100\,Nm^{-1}$$

$$1\,\text{Å}\quad = 100\,pm.$$

For a pure harmonic CO stretch, 2,000 cm^{-1} corresponds to 16·153 md/Å, or 1615·3 Nm^{-1}. The carbon oxygen distances for free CO (r_e) and for CO in $Cr(CO)_6$ (mean, corrected for riding motions) are 112·82 pm (1·1282 Å) and 117·1 pm (1·171 Å). [316, 344]

Note on Figures

The spectra shown in this book were generally taken on a Perkin–Elmer 225 Infrared Spectrophotometer at Glasgow University, under normal operating conditions. The help of Mrs. Frieda Lawrie and of Mr. John Black is gratefully acknowledged.

The cell design and related diagrams loosely follow those of commercially available equipment, principally the Beckman-RIIC cells N-05 (Fig. 4.12), F-05 (Fig. 4.15), and VLT-2 (Fig. 4.16).

CONTENTS

Chapter 1

Introduction

Carbon monoxide is known to form complexes with almost every transition element as well as with the metals of Group IB. The bonding in these complexes can be discussed in terms of two components, a "forward" σ-bond and a "back" or "retrodative" π-bond [1]. The σ-bond is formed by overlap of the highest filled orbital of CO (5σ) with empty orbitals of the metal, while the π-bond is formed by overlap of formally full metal d-orbitals of the correct symmetry with the lowest empty orbital of CO (2π) (Fig. 1.1). Both the 5σ and 2π orbitals are localised mainly on the carbon atom [2] so that one may speak of metal–carbon multiple bonding.† The two components of the bonding are synergic; π-donation by the metal makes it a better σ-acceptor, and causes the CO group to become a better σ-donor. These processes have important chemical consequences, affecting electron and orbital availability at both carbon and metal, stabilising metals in low formal oxidation states, and modifying the bonding and behaviour of other ligands present. Since spectroscopic results can (at least in principle) be interpreted in terms of this model of bonding, the spectroscopy of metal carbonyls is of interest not only to specialists but to inorganic and organometallic chemists in general.

The vibrational spectra of metal carbonyls have proved particularly informative. The CO stretching bands observed in the infrared and Raman spectra are to a good approximation specific group frequencies. They are also sharp, sensitive to environment, and commonly intense. The vibrations of individual MCO groups interact strongly [5], so that the observed spectra are rich in well-resolved bands. The number and pattern of these gives information about molecular symmetry and geometry, while the positions of the bands are related to bonding [6]. The vibrational spectra

† Such bonding was in fact suggested in simpler terms many years before the use of molecular orbital descriptions of complexes became general [3, 4].

1

outside the CO stretching region are also informative, although interpretation is less easy.

A different kind of information can be obtained from the electronic spectra of metal carbonyls, although interpretation is difficult in all but the simplest cases [7]. The results of photoelectron spectroscopy [8, 9] are confined to filled orbitals only, but will hopefully prove more certain. Fairly recently, Fourier transform spectrometers capable of measuring signals due to ^{13}C have become more widely available. Discussions of the carbon-13 n.m.r. spectra of coordinated CO (see e.g. [10, 11]) are still very largely *ad hoc*, but a probe so situated at the atom of greatest chemical interest is likely to attract increasing attention. Nonetheless, the enormous preponderance of vibrational over other kinds of spectroscopic data may be expected to persist, and the arrangement of this book reflects that preponderance.

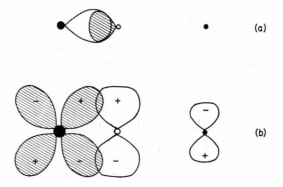

FIG. 1.1. Orbital overlap in metal carbonyl bonding; (a) forward (carbon to metal) bonding, (b) back (metal to carbon) bonding.

Chapter 2 presents a brief account of the formal theory of vibrational spectroscopy for the reader who wishes to understand the theory behind subsequent deductions. The application of this theory to metal carbonyls is discussed in Chapter 3, where the group frequency or "energy factored" treatment of the CO stretching modes in a wide range of possible situations is described in detail. The experimental aspects of vibrational spectroscopy and the extraction of as much information as possible from a spectrum (and no more) are considered in Chapters 4 and 5. Chapter 6 returns to some more difficult topics that are briefly glossed over earlier. The analysis of combination spectra is described, as are the possible implications of anharmonicity; and the presumed relationship between vibrational behaviour and bond type is scrutinised in more detail. In addition, methods used for the assignment of modes other than those due to CO stretching are described and the treatment of spectroscopic data for the solid phase is developed. The bulk of Chapter 7 is, in effect, a literature survey. For obvious reasons (the author noted over two hundred papers in 1971 that presented vibrational data for metal carbonyls) this must be illustrative rather than comprehensive, its purpose being to enable the reader to place new information in its chemical context. In addition, some of the excellent studies that have appeared on a number of individual complexes are discussed in more detail.

Chapter 8 is devoted to some other spectroscopic methods relating to the carbonyl group and the metal electrons that interact with it. Electronic spectra contain d–d bands that place CO in the spectrochemical and nephelauxetic series, and charge transfer bands that can give detailed information about empty orbitals. Low energy photoelectron spectroscopy gives a direct measure of the energy of the metal d electrons, while photoelectron spectroscopy using X-ray sources gives similar information for core electrons on oxygen and carbon, as well as on the metal. Carbon-13 n.m.r. spectroscopy gives chemical shift data for carbon and coupling constants between carbon and other nuclei. The importance of these results is apparent, although their relationship to bonding features at present remains a topic for controversy.

Chapter 2

Principles of Vibrational Spectroscopy

2.1 The Energy of Molecules

The spectra of molecules arise from possible changes in their energy, and the associated emission, absorption, or inelastic scattering of light. Molecules may possess nuclear, electronic, vibrational, rotational and translational energies. For most purposes the interactions between these may be ignored, so that the total energy may be taken simply as the sum of these separate parts. However, additional terms are necessary for molecules in condensed phases, in order to represent interactions with the environment.

The separation of rotational from vibrational energies involves the neglect of several small terms implicit in the dynamics of rotating systems. If a molecule is both rotating and vibrating then the velocity of each nucleus in the molecule will be the sum of parts due to both processes. Since kinetic energy varies as the square of velocity, the expression for the kinetic energy will contain rotation–vibration cross-terms, giving rise to what is known as Coriolis coupling. Further small interactions arise from centrifugal forces in a spinning molecule and from the variation of the molecular moment of inertia with internuclear distance. All these effects are generally too small to be of interest to the inorganic chemist, although Coriolis coupling has on occasion proved of importance in the refinement of force fields.

The separation of vibrational from electronic energies (Born–Oppenheimer approximation) [12] may be thought of in the following way. If the nuclei of a molecule are considered to be in certain definite positions, then the Coulombic energy arising from their mutual electrostatic repulsion is fixed. For a molecule in its ground state,† the electronic energy will be the lowest possible, given the positions of the nuclei. As the nuclei move, so the internuclear repulsion energy changes. At the same time, the electrons

† The argument can readily be extended to other states.

reorganise themselves to produce the lowest electronic energy consistent with the new nuclear positions. The result is that the total vibrational and electronic energy is the sum of three terms. These are the nuclear kinetic energy, the internuclear repulsion energy and the electronic energy. These last two terms provide an effective potential energy in whose field the nuclei move; this is often referred to as *the* potential energy of the molecule. It is assumed that the rearrangement of the electrons (orbital following) is infinitely fast compared with the motion of the nuclei. It is further assumed that the process of orbital following does not cause the electronic energy to depend on the velocities of the nuclei, but only on their positions. The justification for these assumptions lies in the smallness of the electronic mass compared with those of the nuclei.

Three important points can be seen to have emerged from the previous discussion. Firstly, a major component of the effective potential is due to the electrostatic repulsion between the nuclei, and depends only on their separation, rather than on bonding as such. Secondly, the other component can be ascribed to the changes in the total binding energy of the electrons. Thirdly, this latter component is affected by orbital following, as well as by the electronic structure of the molecule at equilibrium internuclear distance. This last point has at times been forgotten in discussions of the significance of vibrational data; it is discussed further in Section 6.4 below.

2.2. Diatomic Molecules

Vibrations of Diatomic Molecules

A diatomic molecule is the simplest system capable of showing molecular vibrations. If two atoms form a stable molecule at all, the variation in energy with internuclear distance will in general be of the type shown in Fig. 2.1. Here it is assumed that the atoms attract each other even at long range so that there is no maximum in the potential energy curve. At very short internuclear distances the forces between the nuclei, and also those between the core electrons of the atoms, give rise to a strong repulsion. Thus there must be some intermediate internuclear distance at which the energy will be at a minimum; this is known as the equilibrium distance.

If q, the deviation of the internuclear distance from its equilibrium value, is sufficiently small, then the potential energy V of the molecule may be expressed in a power series:

$$V = V_0 + aq + kq^2/2 + \text{higher order terms.} \tag{2.1}$$

Potential energy may for convenience be measured from the equilibrium value, so that V_0 vanishes. In addition, the rate of change of V with q is zero when q is zero; thus a also vanishes, and Eq. (2.1) becomes

$$V = kq^2 + \text{higher order terms.} \qquad (2.2)$$

The quantity k is known as the *force constant* for the bond between the two atoms at their equilibrium distance; this is equivalent to the Hooke's law definition of the force constant as the rate of increase of restoring force with displacement. Neglect of the higher order terms in Eq. (2.2) gives the *harmonic oscillator* approximation. Deviations from this approximation have significant consequences which are discussed in Sections 2.2 and 6.1 below. However, the *anharmonicity* caused by such deviations is always small enough to be treated as a perturbation of an oscillator that is very nearly harmonic.

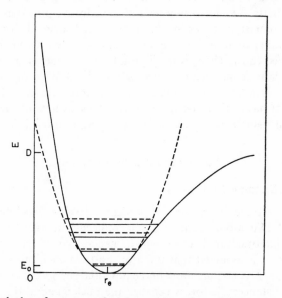

FIG. 2.1. Typical plot of energy against internuclear distance for a diatomic molecule. Dotted line is parabola corresponding to harmonic frequency.

The vibration of a molecule cannot affect the position of its centre of mass. From this it can be shown [13] that the vibration resembles that of a single particle of "reduced mass" μ moving in a field of force constant k. The value of the reduced mass is given by:

$$1/\mu = 1/m_1 + 1/m_2 \qquad (2.3)$$

where m_1, m_2 are the masses of the two nuclei. We should perhaps take m_1,

m_2 to be the masses of the vibrating atoms, rather than of the nuclei alone, to allow for the effects of orbital following, but the distinction is numerically unimportant.

The classical frequency v of oscillation is given by

$$v = \frac{1}{2\pi} \sqrt{\frac{k}{\mu}} \qquad (2.4)$$

There are no classical restrictions on the energy of the molecule, but such restrictions do exist and are consequences of the laws of quantum mechanics. For instance, to assume zero energy would be to define both the positions of the nuclei and their momenta, in violation of the Uncertainty Principle. Standard quantum mechanical arguments give the possible energies of a harmonic oscillator as

$$E = h(v + \tfrac{1}{2}) v \qquad (2.5)$$

where v is an integer, known as the vibrational quantum number, and h is Planck's constant. The possible values of the energy thus form an evenly spaced series. Anharmonicity, however, causes the energy levels of real molecules to deviate from Eq. (2.5); the numerical consequences of this deviation are dealt with in the following section.

Interaction of Diatomic Molecules with Light

Light passing through a molecule imposes an oscillating electric dipole. The variation of this dipole across the molecule may be represented by an oscillating electric quadrupole, while the variation with time implies an oscillating magnetic dipole. These latter effects are both small, and we may regard light as *electric dipole radiation*.

The passage of light through a molecule causes a transient distortion of its electron cloud. This leads to the possibility of elastic (Rayleigh) scattering, important to spectroscopists mainly as a nuisance. More interesting is the possibility of inelastic (Raman) scattering, accompanied by a change in the vibrational quantum number. This possibility arises whenever there is a change in the electronic polarisability of a molecule in the course of a vibration. The frequency v' of the scattered light is then related to v_0, the frequency of the incident light, by the equation

$$v' = v_0 \pm v. \qquad (2.6)$$

Here v is the vibrational frequency of the molecule, as given in Eqs (2.4) and (2.5). The positive sign in Eq. (2.6) corresponds to the possibility of the light gaining energy from the molecule. For this to happen, the process must lead to a decrease in v by one unit, which can occur only where the molecule is not originally in its ground state. The negative sign corresponds to the molecule abstracting energy from the light, with a concomitant increase of one in v. The equation

$$\mathscr{E}' - \mathscr{E}_0 = h(v' - v_0) \tag{2.7}$$

for the energy of the light photon, together with Eq. (2.5), ensures the overall conservation of energy. Since more molecules are in the vibrational ground state ($v = 0$) than in any other, the intensity of the scattered beam of frequency ($v_0 + v$) is less than that of the beam of frequency ($v_0 - v$). Analysis of the inelastically scattered light gives the *Raman spectrum* of the molecule, from which the frequencies of those vibrations causing a change in polarisability can be directly inferred.

Raman spectra are obtained using monochromatic incident light, of frequency far higher than that of the vibrations to be studied. Infrared absorption spectroscopy uses a continuum of incident light including light of the same frequency as these vibrations. If, and only if, a vibration involves a change in the molecular dipole moment, then it is possible for light of frequency v to be absorbed, the light energy being transformed into vibrational energy of the molecule. For diatomic molecules in particular, infrared absorption can occur whenever the two atoms are different. The intensity of the absorption depends, among other things, on the square of the oscillating molecular dipole. Two factors contribute to the size of this dipole. They are the charges on the moving atoms, and the orbital following of the valence electrons. The former of these is responsible for high infrared intensities in the spectra of polar species; the latter, as discussed in Chapter 6, is responsible for the strength of the absorption spectra of metal carbonyls. Molecules in excited states can, in certain circumstances, emit light, but this property has not been exploited in vibrational spectroscopy.

The selection rule for the absorption, emission, or scattering of electric dipole radiation by a simple harmonic oscillator is

$$\Delta v = \pm 1 \tag{2.8}$$

so that for such an oscillator only one frequency should be observable. From the masses of the atoms it is easy to calculate the force constant. If the masses of the atoms are expressed in atomic units, the frequency of the light in cm^{-1}, and the force constant in Newtons metre^{-1} (Nm^{-1}),

$$k = 0 \cdot 58915 \times 10^{-4} \, v^2 \, \mu. \tag{2.9}$$

In particular, if the atoms concerned are ^{12}C and ^{16}O, then

$$k(CO) = 4 \cdot 0383 \times 10^{-4} \, v^2. \tag{2.10}$$

The anharmonicity of real molecules has two effects. The selection rule of

Eq. (2.8) does not apply rigorously to real molecules. As a result, it is possible to observe *overtones*, in which v changes by two or more units.†
In addition, the frequencies deviate to some small extent from those required by Eqs (2.4, 2.5, 2.9). Equation (2.5) becomes to a good approximation (but see Section 6.2)

$$E = E_0 + \sum_k \omega_k (v_k + \tfrac{1}{2}) + c \sum_{k \leqslant l} X_{kl} (v_k + \tfrac{1}{2}) (v_l + \tfrac{1}{2}) \tag{2.11}$$

often written in the form

$$E = E_0' + \sum_k \omega_k' v_k + c \sum_{k \leqslant l} X_{kl}' v_k v_l \tag{2.12}$$

where the sum runs over all modes.‡

The frequency ω is that which the molecule would show if the higher order terms of Eq. (2.1) were strictly zero. It may be found using Eqs (2.11, 2.12) if overtones are observed, and should strictly be used in Eqs (2.9, 2.10) in place of v. Higher order terms exist, but may be ignored for our purposes.

2.3. Polyatomic Molecules

Any account of the vibrations of a polyatomic molecule is bound, for several reasons, to be complex. To specify the positions of n atoms, $3n$ Cartesian coordinates are required. However, since a molecule generally possesses six degrees of translational and rotational freedom (five for a linear molecule), the shape of the molecule may in fact be specified by $3n - 6$ (or $3n - 5$) *internal coordinates*, such as angles and bond lengths. While the kinetic energy of the vibrating atoms is most readily expressed using the Cartesian coordinates, it is the distortion of the internal coordinates that determines the potential energy. This complicates the formulation of useable equations of motion, and the complexity of this task is increased by the presence in the molecule of atoms of different masses. Finally, the force field that describes the resistance of the molecule to distortion contains a large number of terms. Not only is a term required for each individual internal coordinate, but interaction terms are also needed, because altering one coordinate can significantly affect the equilibrium value of the others.

---

† For molecules with more than two atoms, anharmonicity also makes possible the observation of *combination bands*. These arise from concerted changes in the quantum numbers of two or more different vibrational modes.

‡ Including all individual modes of degenerate representations. Alternatively, the sums can be taken to include one term only from each degenerate set, the vibrational quantum number of which is taken as the sum of the vibrational quantum numbers of its components. It is then necessary for each degenerate set i to include a term $G_{ii}l_i^2$, and other, related, terms, where l is the internal rotational quantum number for the internal mode, defined in Appendix XIV of Reference [13].

The formal theory required to handle these complexities is well developed, and is described in the following section. The account given, *which may be omitted on first reading*, is meant to show the relationship between the formal theoretical apparatus and simpler concepts. The succeeding Section summarises the main results of the more formal analysis and gives a qualitative account of the significance of the concepts used.

Formal Treatment

Normal coordinates and normal modes. The vibrations of a polyatomic molecule may (if anharmonicity is neglected) be expressed in terms of the behaviour of a limited number of normal modes. Each of these is an oscillation of a normal coordinate of the molecule and the normal coordinates can conveniently be written as linear combinations of mass-weighted coordinates. The mass-weighted coordinates $q_1, q_2 \ldots q_{3n}$ for a molecule with n constituent atoms are defined by

$$q_1 = \sqrt{m_1} \, \Delta x_1$$
$$q_2 = \sqrt{m_1} \, \Delta y_1$$
$$q_3 = \sqrt{m_1} \, \Delta z_1$$
$$q_4 = \sqrt{m_2} \, \Delta x_2 \text{ etc.} \qquad (2.13)$$

The kinetic energy T of the molecule is given by

$$T = \tfrac{1}{2} \sum_i \dot{q}_i^2. \qquad (2.14)$$

The potential energy V may be expanded in a power series in the q_i, and v_0 and terms linear in q_i eliminated, as for a diatomic molecule. This gives

$$V = \tfrac{1}{2} \sum_i \sum_j f_{ij} q_i q_j. \qquad (2.15)$$

For any set of particles, Newton's equations of motion may be written as the set of simultaneous equations

$$\frac{d}{dt} \frac{\partial T}{\partial \dot{q}_i} = -\frac{\partial V}{\partial q_i}. \qquad (2.16)$$

In the elementary terms, the left-hand side of this equation may be thought of as giving the rate of change with time of momentum along the coordinate q_i, while the right-hand side is regarded as the force responsible for that change. Combining Eqs (2.14–2.16) gives

$$\ddot{q}_i + \sum_j f_{ij} q_j = 0 \,\dagger \qquad (2.17)$$

It is possible to choose new normal coordinates, Q, which are linear combinations of the mass-weighted coordinates, so that the expressions for V *and* T can be written without cross-terms:

$$q_i = \sum_k B_{ik} Q_k, \qquad (2.18)$$

$$T = \tfrac{1}{2} \sum_k \dot{Q}_k, \qquad (2.19)$$

$$V = \tfrac{1}{2} \sum_k \lambda_k Q_k^2. \qquad (2.20)$$

The equation corresponding to Eq. (2.17) is then

$$\ddot{Q}_k + \lambda_k Q_k = 0 \qquad (k = 1, 2 \ldots 3n). \qquad (2.21)$$

This is the differential equation for a harmonic oscillation in the coordinate Q_k, with frequency $\sqrt{\lambda_k}/2\pi$, of arbitrary amplitude and phase

$$Q_k = A_k \sin\left(\sqrt{\lambda_k}\, t - \phi_k\right). \qquad (2.22)$$

Consider an oscillation in which one particular Q_k vibrates with unit amplitude, while the other normal modes do not vibrate. Then

$$q_i = B_{ik} Q_k = B_{ik} \sin\left(\sqrt{\lambda_k}\, t - \phi_k\right). \qquad (2.23)$$

Substituting from Eq. (2.23) into Eq. (2.17), the q's may be eliminated to give

$$- \lambda_k B_{ik} + \sum_j f_{ij} B_{ik} = 0 \qquad (2.24)$$

for each value of i. This is a set of $3n$ simultaneous equations for the coefficients B_{ik}, and the condition for any solution to exist (other than the trivial solution in which all the B_{ik} are zero) is

$$|f_{ij} - \delta_{ij} \lambda_k| = 0. \qquad (2.25)$$

Here δ_{ij} is the Kroneker delta function, equal to one where $i = j$ but zero where i, j differ. Thus λ_k is a solution of the secular equation

$$|f_{ij} - \delta_{ij}\lambda| = 0. \qquad (2.26)$$

† In the derivation of Eq. (2.17), the factor of $\tfrac{1}{2}$ associated with $f_{ij} q_j (i \neq j)$ vanishes by summation, since the two-dimensional sum of Eq. (2.15) contains both $f_{ij} q_i q_j$ and $f_{ji} q_j q_i$. Moreover,

$$f_{ij} = \partial^2 V / \partial q_i \partial q_j = \partial^2 V / \partial q_j \partial q_i = f_{ji}$$

Equation (2.26) is of order $3n$, and refers to all motions of the molecule including rotations and translations. Thus there are in general six "normal coordinates" that leave the molecule undistorted, and for which λ is zero. If the quantities f_{ij} of Eq. (2.15) were known, it would then be possible to calculate the remaining $3n - 6$ possible values of λ, and hence the frequencies of the normal modes of vibration. Substituting for λ in the set of Eqs (2.24), would then give the relative amplitudes $B_{1k} : B_{2k} : B_{3k} \ldots B_{3nk}$ for the relative contributions of each of the mass-weighted coordinates to the kth normal mode.

General coordinates. The derivation of the secular equation may be carried out for any set of independent coordinates r_i sufficient to describe the system. The expressions for the kinetic and potential energy are then of the form

$$T = \tfrac{1}{2} \sum_i \sum_j t_{ij}\, \dot{r}_i \dot{r}_j, \tag{2.27}$$

$$V = \tfrac{1}{2} \sum_i \sum_j v_{ij}\, r_i r_j. \tag{2.28}$$

Strictly speaking, the t_{ij} may themselves be functions of the r_i. For instance, if bond lengths and angles are used as coordinates, the kinetic energy of an atom A due to bending of an angle ABC depends on the length of the AB bond. Such effects may, however, be neglected if the displacements from equilibrium are small. With

$$r_i = \sum_k C_{ik} Q_k, \tag{2.29}$$

then

$$- \lambda_k \sum_j t_{ij}\, C_{jk} + \sum_i v_{ij}\, C_{jk} = 0. \tag{2.30}$$

$$|v_{ij} - t_{ij}\lambda| = 0, \tag{2.31}$$

corresponding to Eqs (2.24, 2.26) respectively. The development of techniques for finding the v_{ij}, t_{ij}, and for defining coordinates that facilitate this task, are then central aims of the theory.

Internal coordinates and symmetry coordinates. The distortion of a molecule from equilibrium can be written in terms of the internal coordinates s, which are the bond lengths, bond angles, bond-plane angles and dihedral angles. There are two obvious advantages in doing this. Firstly, the elements of Eq. (2.28) are then equal, by definition, to the force constants F_{ii} and interaction constants F_{ij} of a generalised valence force field. Secondly, by a judicious choice of coordinates it is possible to eliminate the overall translations and rotations of the molecule, reducing the order of the secular determinant from $3n$ to $3n - 6$. There is a further, more subtle advantage. It is particularly easy to combine the internal coordinates into linear combinations that span irreducible representations of the molecular point group. Since species belonging to different irreducible representations cannot interact, the effect of this procedure is to factorise the secular determinant into a number of determinants of lower order, one for each irreducible representation spanned.

Since the potential energy may readily be defined in terms of internal coordinates (or symmetry coordinates constructed from these), it only remains to find an expression for the kinetic energy in order to formulate the secular equation. The G-matrix method, described below, gives a convenient procedure for doing this.

The **F** *and* **G** *matrices.* The internal coordinates s may be expressed in terms of the Cartesian coordinates u_j:

$$s_k = \sum_j D_{kj} u_j. \tag{2.32}$$

The elements of the **G** matrix are then defined by

$$G_{kl} = \sum_i (D_{ki} D_{li}/m_i) \tag{2.33}$$

where the summation is carried out over all the Cartesian coordinates u_i, and m_i is the mass of the atom to which u_i applies. It can be shown [13] that (2.31) then becomes

$$|\mathbf{F} - (\mathbf{G})^{-1} \lambda| = 0, \tag{2.34}$$

or

$$|\mathbf{FG} - \lambda \mathbf{E}| = 0, \tag{2.35}$$

or

$$|\mathbf{GF} - \lambda \mathbf{E}| = 0. \tag{2.36}$$

Thus the off-diagonal elements in λ have been removed, while the diagonal elements occur with coefficients of one.

F and **G** matrix elements for a number of species of type $L_x M(CO)_n$, are given in Appendix II.

Discussion of Formal Treatment

A diatomic molecule vibrates as a simple harmonic oscillator. The effective mass μ of this oscillator is given by Eq. (2.3). The restoring force operating on this effective mass is the force constant k, and the classical motion of the oscillator is given by

$$s = A \sin (\sqrt{\lambda} t - \phi) \tag{2.37}$$

where

$$k \mu^{-1} - \lambda = 0. \tag{2.38}$$

A polyatomic molecule may be thought of as a set of $3n - 6$ coupled oscillators ($3n - 5$ if the molecule is linear). These correspond to the $3n - 6$ internal degrees of freedom, as shown by the internal coordinates. Coupling between the oscillators arises in two distinct ways, giving rise to what we shall call "dynamic" and "kinematic" coupling. Dynamic coupling arises because of the effects of one internal coordinate on the equilibrium values of the others; it is responsible for the interaction constants in the expression for potential energy. Kinematic coupling arises because the distortion of one internal coordinate involves the movement of atoms which are also associated with other internal coordinates. The generalisation of Eq. (2.38) to a polyatomic molecule is Eq. (2.34), and the generalisations of force constant and effective mass are the F and $(G)^{-1}$ matrices.

Equation (2.34) can be re-expressed as Eq. (2.35) or as Eq. (2.36), thus eliminating off-diagonal elements in λ. The F matrix is constructed directly from the force constants and interaction constants. The meaning of the G matrix, which is defined by Eqs (2.32, 2.33), is less obvious, but it can be shown to be a generalisation of reciprocal effective mass. The detailed procedure for finding the G matrix has been given, as have the G-matrix terms arising from a large number of atomic arrangements (Ref. 13, Section 4.2 and Appendices VI, VII).

Equation (2.35) can in many cases be greatly simplified by using the symmetry properties of the molecule to which it is applied. This is done by combining the internal coordinates into symmetry coordinates, which span irreducible representations of the molecular point-group. The secular determinant is thereby automatically factorised, and a separate secular equation may be written for each irreducible representation spanned.

Simplified F matrices and symmetry coordinates for the high-frequency carbonyl group frequencies are presented in the following chapter.

Determination of Force Constants from Frequencies

The theory given so far would enable us to predict the frequencies of molecules if we knew their force fields. In fact, the problem is always the other way round; we observe frequencies and use them to estimate force fields. The first step in this procedure must be to assign the observed frequencies to the correct normal modes, i.e. to the correct symmetry species. This is done by a combination of techniques discussed throughout the book. These techniques include the use of approximate force fields, the collection of intensity data, and the application of selection rules for fundamental and also for combination bands, comparison of related species and the establishment of detailed trends, isotopic substitution, polarisation data, the comparison of solution with solid state and vapour phase spectra, and the study of solvent effects.

For each symmetry species, the observed frequencies v give the possible values of λ:

$$\lambda_i = 5\cdot8915 \times 10^{-5} v_i^2. \tag{2.39}$$

Substituting each of the λ_i in turn into Eq. (2.35) then gives a series of simultaneous equations in the elements of the $\mathbf{F}$ matrix. An equivalent but less unwieldy series can be found by equating powers of λ in the equation

$$|\mathbf{F G} - \lambda\mathbf{E}| = (\lambda_1 - \lambda)(\lambda_2 - \lambda) \ldots (\lambda_i - \lambda) \ldots . \tag{2.40}$$

If Eq. (2.40) is of order m in λ, then equating coefficients of $\lambda^0, \lambda^1, \ldots \lambda^{m-1}$ gives m simultaneous equations. However, the $\mathbf{F}$ matrix contains $m(m + 1)/2$ independent elements. Thus (except where $m = 1$ for all representations spanned) we have an underdetermined problem, with more unknowns than equations, so that there is not enough information to specify a solution.

The amount of information may be increased by isotopic labelling; ^{13}CO and $C^{18}O$ have both been widely used for this purpose. The number of unknowns may be reduced by assuming that certain interaction constants are zero, that particular relationships hold between some of the constants, or that certain force constants may be transferred from one molecule to another.

The number of unknowns, and the difficulty of finding them, may be drastically reduced by using the approximation of group frequencies, sometimes known as the "energy factoring approximation". In this approximation, sets of vibrations are assigned exclusively to particular sets of functional groups of the molecule. All elements of the $\mathbf{F}$ or $\mathbf{G}$ matrices that link different sets of functional groups are simply ignored. In metal carbonyls, for instance, all bands around 2000 cm^{-1} are assigned to pure CO stretching modes. Few would claim absolute validity for the stretching parameters (one could hardly call them force constants) derived in this way; but it seems reasonable to hope that the trends shown by these parameters may parallel trends in the force constants to which they approximate.

After all acceptable† approximations have been made, and after all accessible† information has been gathered, the problem may still remain underdetermined. If there is only one degree of freedom, i.e. one more unknown than there are independent equations, then it is possible to express all the unknowns in terms of a single parameter, and thus to display graphically all the possible sets of solutions. If the number of degrees of freedom exceeds one, then the possibilities proliferate uncontrollably.

† These terms are of course somewhat subjective. What counts as an acceptable approximation depends on the degree of sophistication required; and what counts as accessible information depends on the lengths to which the investigator is prepared to go. The number of different isotopically labelled species that can be examined, for instance, is quite large.

On occasion, the problem may be overdetermined, in which case a set of force constants can be found by discarding some of the data. Experimental errors and the effects of approximations will, however, make it unlikely that any set of force constants can be found to fit all the data exactly. The force constants found will therefore depend on which data are discarded. This is not a satisfactory situation and it is preferable to find a set of force constants that gives a best fit to all the data. Fortunately, computer programs which do this have been developed [14].

"Best fit" programs may also be used instead of algebraic manipulation to solve determinate problems. Indeed, sometimes attempts to solve determinate problems algebraically give imaginary roots, and a best fit is as much as can be hoped for. This is often the situation when approximate parameters are being determined for substituted metal carbonyls, especially if assumed relations between the interaction constants are used. In this case it is possible for an erroneous assignment to lead to plausible real parameters, while the correct assignment, firmly established from IR and Raman intensity data, leads only to imaginary solutions.

Except where m equals one for all representations, finding force constants must involve the solving of an equation of order two or higher. Such equations generally admit more than one solution, so that even for determinate problems several different alternative sets of force constants could explain the same data. This ambiguity is rarely serious, since many mathematically acceptable solutions may be rejected as chemically unreasonable.

Redundant Coordinates and Inaccessible Force Constants

It was stated above that symmetry coordinates can be found by taking linear combinations of internal coordinates. Occasionally such linear combinations correspond to impossible vibrations, or to overall translations and rotations. This situation is readily illustrated by the in-plane bending motions of a planar molecule AB_x. For such a molecule it is necessarily the case that the angles BAB add up to $360°$. The fully symmetric combination of BAB bends describes a process that violates this condition. It is always the case that G-matrix elements involving such processes are zero. The symmetric sum of angular displacements is said to be a redundant coordinate, and the geometric condition that the sum of the angular displacements be zero is termed a redundant condition. The redundant coordinate should simply be discarded.

In some cases, a symmetry coordinate can describe an overall rotation. For example, the out-of-plane terminal atom displacements of a square planar molecule AX_4 span the representations A_{1u}, B_{2u}, E_g of the point-group D_{4h}. The coordinates of symmetry A_{1u}, B_{2u} represent possible distortions.

Those of symmetry E_g represent processes in which one X atom is displaced upwards while that *trans* to it is displaced downwards. This process is not a distortion at all but a rotation of the molecule around an axis through the other two X atoms. In this case, all $\mathbf{F}$ matrix elements involving the coordinate are zero.

Closely related to redundant coordinates are inaccessible force constants. These may be illustrated by the in-plane bending modes of a planar species AXY_2 of symmetry C_{2v}. Let the angles XAY be α, and the angle YAY be β. Then the symmetric expansion of the two angles α gives an $\mathbf{F}$-matrix element

$$F_{\alpha\alpha} + 2F_{\beta\beta} + F_{\alpha\alpha'} - 4F_{\alpha\beta}.$$

The asymmetric expansion of one α while the other contracts, gives an element

$$F_{\alpha\alpha} - F_{\alpha\alpha'}.$$

(In these expressions the $F_{\alpha\alpha}$, $F_{\beta\beta}$ are diagonal elements corresponding to the distortion of the angles, while $F_{\alpha\alpha'}$, $F_{\alpha\beta}$ are interaction constants.) It is possible to find two combinations of the four parameters involved, but it is not possible to find any of the parameters separately. Thus $F_{\alpha\alpha}$ and $F_{\beta\beta}$ cannot both be determined. This is a consequence of molecular geometry, and no amount of experimental evidence can make the problem determinate. A purely formal solution can be imposed by putting $F_{\alpha\beta}$ and $F_{\beta\beta}$ equal to zero, but the significance for bonding of the force constants so obtained is not clear.

Redundant coordinates and inaccessible force constants can be constructed from linear as well as from angular displacements. The distance between A and X in a linear molecule of type ABX is one such redundant coordinate, since it may be defined by the distances between A and B, and between B and X. This redundant condition has an important physical consequence. Should there exist forces between A and X, they cannot be distinguished from the forces between A and B, and between B and X. Thus there is no possibility of detecting forces between metal and oxygen in metal carbonyls.

Relative Amplitudes and Potential Energy Distribution

The approximation of group frequencies assumes that only one kind of internal coordinate is affected by a given normal mode, but this is always an over-simplification. It is thus of interest to know how much each of the symmetry coordinates contributes to the actual vibrations. There are two measures of such contributions, the relative amplitudes of the vibrations of each symmetry coordinate, and the potential energy distribution.

The symmetry coordinates may be expressed in terms of the normal coordinates

$$S_i = \sum_k L_{ik} Q_k \tag{2.41}$$

where, for every value of i,

$$\sum_j [(\mathbf{GF})_{ij} - \delta_{ij} \lambda_k] L_{jk} = 0 \tag{2.42}$$

and the elements $(\mathbf{GF})_{ij}$ are defined by the matrix multiplication rule

$$(GF)_{ij} = \sum_t G_{it} F_{tk}. \tag{2.43}$$

Here Eqs (2.41, 2.42) are special cases of Eqs (2.29, 2.30). The ratio

$$L_{1k} : L_{2k} : \ldots L_{1k} \ldots L_{mk}$$

then expresses the relative amplitudes of vibration of the symmetry coordinates $S_1, S_2 \ldots S_2 \ldots$ in the vibration of the normal mode Q_k.

The potential energy distribution of a vibration may be expressed by substituting from (2.41) into the equation

$$V = \tfrac{1}{2} \sum_i \sum_j F_{ij} S_i S_j \tag{2.44}$$

which is a special case of (2.28). Since cross-terms in Q_i, Q_j $(i \neq j)$ vanish, it follows that

$$V = \tfrac{1}{2} \sum_k \left(\sum_i \sum_j F_{ij} L_{ik} L_{jk} \right) Q_k^2. \tag{2.45}$$

The terms in (2.45) where $i = j$ are ascribed to the individual symmetry coordinates, while the terms for $i \neq j$ are ascribed to their interactions. The ratio

$$F_{11} L_{1k}{}^2 : F_{12} L_{1k} L_{2k} : \ldots F_{ij} L_{ij} L_{jk} \ldots F_{mm} L_{mk}^2$$

then expresses the relative contributions of each of the symmetry coordinates and of their interactions to the potential energy of vibration of the kth normal mode.

It is often the case that one of the L_{ik} is much larger than the others, so that the kth normal mode is predominantly a vibration of one particular internal coordinate. The group frequency approximation then provides a meaningful qualitative description of the nature of the vibration. In some cases, however, a vibration can only be discussed as a mixture of distortions of two or more internal coordinates.

Compliance Constants and Interaction Coordinates

Although the most usual way of describing the force field of a molecule is by the use of force constants, it is not the only way. One alternative is the use of the compliance constant matrix, defined as the inverse of the force constant matrix

$$C = (F)^{-1}. \tag{2.46}$$

The diagonal elements C_{ii} are known as compliance constants of the s_i. It can be shown that the compliance constant for a given internal coordinate is equal to the displacement suffered by that coordinate under the influence of a unit force, *provided no forces are exerted on the other internal coordinates*. This proviso implies that in general the other internal coordinates s_j $(i \neq j)$ will be displaced so as to minimise the potential energy in the new force field created by the distortion of s_i. The displacement of s_j when unit force is applied to s_i is equal to C_{ij}.

The main advantages of compliance constants are that they are uniquely defined, that they are determinate (even for those coordinates involved in redundancies), and that they simplify the treatment of centrifugal distortion. The main disadvantages are unfamiliarity, and a less direct relationship with the potential energy surfaces for distortions. Despite these disadvantages, it seems probable that compliance constant data will find increasing use in sophisticated treatments of metal carbonyls.

2.4. Selection Rules and Intensities

In Section 2.2 it was stated that (within the electric dipole approximation for light) those vibrations of a diatomic molecule that give rise to a change in dipole moment are infrared-active, while those that give rise to a change in polarisability are Raman-active. These statements are also true for polyatomic molecules and may be extended to crystalline arrays.

In the discussion of infrared or Raman spectra, vibrations may be classified as "observable" or "unobservable". This distinction is based on experiment, and can become unclear when bands are very weak. A more fundamental distinction is that between "allowed" and "forbidden" bands. A band is classified as infrared-allowed if it is due to a vibration that is allowed by the symmetry of the molecule to give rise to an oscillating dipole. The vibration is then classified as infrared-active. Likewise, a Raman-allowed band arises from a vibration allowed to give rise to a change in polarisability. The classification of bands as "allowed" and "forbidden" depends on the assumption of a molecular point-group; this assumption may involve a certain degree of idealisation. The fact that a band is allowed does not imply that it will have any measurable intensity, and it is possible for allowed bands to be vanishingly weak. It is also possible for "forbidden" bands (classified as such on the basis of an over-idealised geometry) to be readily observable.

Selection Rules for Infrared Spectra

As stated above, a vibration is infrared-active if, and only if, symmetry permits that vibration to give rise to an oscillating electric dipole. For this to be the case, the vibration must belong to the same irreducible representation of the molecular point-group as does a displacement along some axis. The vibration is then allowed to absorb light of which the electric vector is polarised parallel to this axis. To find out whether a vibration is active or inactive, all that is necessary is to see whether it belongs to the same irreducible representation as x, y or z; group-theoretical character tables generally display this information.

For polyatomic molecules of less than cubic symmetry, some vibrations are allowed for light polarised along the conventional molecular z-axis, while others are polarised along the x and y axes. For example, the symmetric (a_1) and asymmetric (e) CO stretching modes of species cis-$L_3M(CO)_3$ are z- and x, y-polarised respectively. The direction of polarisation of infrared absorption has no implications for solution spectra, since the axes of the molecules in normal solutions are arranged at random. Studies with polarised light can, however, be used for solutes oriented in a nematic solvent [15], and are extremely valuable in the study of solids. In some cases it is possible to distinguish between z- and x, y-polarised absorption from the forms of the bands observed for vapour phase samples, since the selection rules governing the changes in rotational quantum number for these are different [13, Appendix XVI]. The usefulness of rotational structure as an aid to assignment in metal carbonyl chemistry is, however, rather limited. The number of volatile metal carbonyl derivatives that remain to be investigated must be quite small; and the moments of inertia of all but the simplest metal carbonyl derivatives are large.

The Intensities of Infrared Bands

If the allowedness of an infrared vibration depends on its giving rise to an oscillating dipole, its intensity depends on the square of the size of that dipole. A large oscillating dipole can arise either from the permanent polarity of a bond, or from orbital following. Vibrations involving a highly ionic bond give rise to a large oscillating dipole by altering the charge separation. This is not the situation for carbon monoxide, nor for metal carbonyl complexes, since the polarity of the bonds in these species is low [2, 16]. Despite this, the infrared bands associated with the CO stretching motions of metal carbonyls are generally intense. This intensity is a consequence of the peculiar nature of the metal–carbonyl bond, which leads to an unusually high degree of orbital following.

The bonding in metal carbonyls has been much discussed [16, 17, 18], and the implications for bonding of vibrational spectra are considered further below (Section 6.4). It is generally accepted that the metal–carbonyl bond involves extensive π-overlap between filled d-orbitals on the metal on the one hand, and the empty 2π (C—O antibonding) orbital of CO on the other. The extent to which CO 2π accepts charge will depend on the relative energies of the orbitals involved. Stretching a C—O bond lowers the energy of 2π by reducing the magnitude of the antibonding interaction between carbon and oxygen, and may thus be expected to lead to an increase in the amount of metal to carbonyl π-donation. This effect is liable to be significant, since the classical amplitude of a carbonyl ligand in its first excited vibrational state is 6 pm, which corresponds to a lowering of the 2π energy by 20,000 cm^{-1} [19]. It is the high degree of orbital following in the metal–carbon bond, consequent on the stretching of a C—O bond, that is responsible for the high infrared intensities generally observed.

One possible complication requires special comment. If the form of a vibration is such that the oscillating dipoles due to the individual carbonyl groups involved are opposed, then the infrared band due to that vibration is liable to be weak. The limiting case is that of a symmetry-forbidden band, the intensity of which is ideally zero. It is, however, possible for even a symmetry-allowed band to be so weak as to cause difficulties in assignment [20, 21].

Bands due to metal–carbon stretching modes are weaker than those due to C—O stretching, but generally stronger than those due to MCO bending. The implication is that alterations in metal–carbon bond length have a smaller effect on the nature of the metal–carbon bond than changes in the distance between carbon and oxygen.

Selection Rules for Raman Spectra

The elementary treatment of Raman spectra [13, Sections 3.6, 3.7] shows that a molecular vibration is Raman-active if and only if the polarisability α of the molecule changes during the vibration. Polarisability connects two vector quantities; an external electric field imposed on a molecule, and the dipole that this field induces. It is therefore a tensor quantity, with independent components α_{xx}, α_{yy}, α_{zz}, α_{xy}, α_{yz}, α_{zx} ($\alpha_{yx} = \alpha_{xy}$ etc). Here α_{xx} gives the dipole induced along the chosen x-axis of the molecule by an electric field also oriented along that axis, while α_{xy} gives the y-axis dipole due to an x-axis field.

The symmetry condition for a vibration to be Raman-active is then that it must belong to the same irreducible representation as do any of

x^2, y^2, z^2, xy, yz or zx. The relevant irreducible representations may be found by standard group-theoretical methods, and are generally included in character tables. In particular, the quantity $(\alpha_{xx} + \alpha_{yy} + \alpha_{zz})/3$, known as the mean polarisability, always belongs to the fully symmetric representation. It follows that vibrations belonging to this representation are always Raman-allowed.

In many point-groups it is possible for vibrations to be infrared-active, or Raman-active, or both, or neither. For instance, in the point-group D_{3h}, vibrations of symmetry A_1' and E'' are Raman-active but infrared-inactive, those of symmetry A_2'' are infrared-active but Raman-inactive, those of symmetry E' are active in both Raman and infrared spectra, and those of symmetries A_1'' and A_2' are active in neither. For molecules with a centre of symmetry, it is not possible for a vibration to be active in both kinds of spectrum, since all infrared-active vibrations are odd with respect to inversion and all Raman-active vibrations even. This principle of mutual exclusion also applies to some point-groups lacking a centre of symmetry, including the point-group D_{4d} to which dimanganese decacarbonyl belongs; it is, however, far from universal.

Depolarisation Ratios†

The infrared bands of disordered solutions give no direct information about the symmetries of the vibrations that cause them. Raman spectra, however, can distinguish between fully symmetrical bands and others by examination of the degree of polarisation of the scattered light.

In Raman scattering, a molecule is excited by electric dipole radiation of a certain frequency and emits electric dipole radiation that differs in frequency and may also differ in polarisation. The dipoles of the exciting and scattered radiation are connected by a second rank tensor $\mathbf{R}$; this is characteristic of the molecular vibration responsible for the scattering and belongs to the same representation of the molecular point group as that vibration.

In solution, molecules adopt all possible orientations. The effect of this is to average the $\mathbf{R}$ tensor, and it can be shown that the averaged scattering tensor $\bar{\mathbf{R}}$ is given by

$$\bar{R}_{ii}^2 = \bar{R}_{xx}^2 = \bar{R}_{yy}^2 = \bar{R}_{zz}^2 = [45\,\bar{R}^2 + 4\gamma^2\,(R)]/45, \qquad (2.47)$$

$$\bar{R}_{ij}^2 = \bar{R}_{xy}^2 = \bar{R}_{yz}^2 = R_{zx}^2 \text{ etc} = \gamma^2\,(R)/15, \qquad (2.48)$$

where $\bar{R}$ is the mean value of the tensor $\mathbf{R}$

$$\bar{R} = (R_{xx} + R_{yy} + R_{zz})/3 \qquad (2.49)$$

† The notation used here follows that of [22, Section 3.1], which differs in detail from that of [13, Sections 3.6, 3.7].

and $\gamma(R)$ is the anisotropy

$$\gamma^2(R) = (R_{xx} - R_{yy})^2 + (R_{yy} - R_{zz})^2 + (R_{zz} - R_{xx})^2$$
$$+ 6(R_{xy} + R_{yz} + R_{zx})/2 \qquad (2.50)$$

Consider the scattering of exciting light polarised along the x-axis and propagated along the z-axis. Scattered light collected in the yz plane will have two components, polarised parallel and perpendicular to x. The intensity ratio ρ_l of perpendicular to parallel polarised light is then given by

$$\rho_l = I_{\perp x}/I_{\parallel x} = \bar{R}_{ij}^2/\bar{R}_{ii}^2$$
$$= 3\gamma^2(R)/[45\bar{R}^2 + 4\gamma^2(R)],$$

since the intensity of light is proportional to square of the oscillating electric dipole. The quantity ρ_l is known as the depolarisation ratio for linearly (i.e. plane) polarised light.

It follows from the definition (2.49) of the mean value $\bar{R}$ that it belongs to the fully symmetric representation of the molecular point group. But it is also true that $\bar{R}$ belongs to the same representation as the vibration responsible for the scattering. Thus, unless this vibration belongs to the fully symmetric representation, $\bar{R}$ is zero and ρ_l takes the value 3/4. For a fully symmetric vibration, $\bar{R}$ is positive, and generally not zero, so that the depolarisation ratio is less than 3/4. A depolarisation ratio of less than 3/4 is therefore enough to identify a band as being due to a fully symmetric representation. The converse, of course, is not true. It is possible for a vibration to be fully symmetric, and yet to give a depolarisation ratio not measurably less than 3/4, if $\bar{R}$ is very much smaller than $\gamma(R)$.

The argument given above for the scattering of plane polarised light can also be applied to non-polarised (natural) light. The incident beam may conveniently be regarded as the sum of two plane polarised beams. The scattering of each of these may be treated separately. The value of ρ_n, the depolarisation ratio for natural light observed at right angles to the direction of propagation, is then found to be between zero and 6/7 for fully symmetric vibrations, and equal to 6/7 for all others. The additional information that may be obtained for crystalline samples, is discussed further in Chapter 6.

Chapter 3

The CO Stretching Modes of Metal Carbonyls

3.1. General

Metal complexes containing one or more CO ligands bound to a single metal atom show one or more intense infrared bands between 2200 and 1800 cm^{-1}, which are assigned to the carbon–oxygen stretch of the coordinated CO. The CO stretching motions are highly sensitive to environment, and exhibit large interactions. As a result the band frequencies give information about the electron availability within the molecule, while their number, general appearance, and relative intensity give information about the structure. The study of these bands is thus highly rewarding.

In this Chapter, the carbonyl stretching vibrations of a range of commonly found arrangements of general type $L_xM(CO)_y$ are discussed. Energy factored force fields are presented; and some approximations often imposed on these already incomplete force fields are described. The treatment is further extended to polynuclear metal carbonyls.

But first it is necessary to evaluate the energy factoring procedure itself. This is done here by detailed analysis of the simplest model systems MCO and $M(CO)_2$.

3.2. The Vibrations of an Isolated MCO Group

In this section we discuss the vibrations of a hypothetical isolated MCO fragment with a particular set of force constants, from which the frequencies may be calculated. The problem of finding the stretching force constants from the frequencies is, however, underdetermined; but if a value is assumed for the interaction constant $F_{MC, CO}$ the remaining force constants may be calculated, as can the frequencies of $M^{13}CO$ and $M^{12}C^{18}O$.

This discussion serves several purposes. It illustrates for the simplest possible case the formal theory of the preceding Chapter. It also makes possible a discussion of the relationship between true force constants and energy-factored parameters, illustrates the use of isotopic labelling in the assignment of bands and in the calculation of force constants, and reveals the problems that arise in attempting to fit data for isotopically labelled species to energy factored force fields.

For the stretching force constants F_{MC} and F_{CO}, the bending force constant F_β, and the interaction constant $F_{MC, CO}$, we assume values of 200 Nm^{-1}, 1700 Nm^{-1}, 4·6 × 10^{-19} Nm rad^{-2}, and 70 Nm^{-1} respectively. These values are close to those found in detailed treatments of symmetrical metal carbonyls [23]. For concreteness, we assume a mass number of 60 for M, and MC and CO distances of 192 and 117 pm respectively.

MCO Bending Modes

The two bending modes of linear MCO are degenerate, belonging to the representation Π of the point group $C_{\infty v}$, while the stretching modes belong to the representation Σ. It follows immediately that, in this model system, stretching and bending modes cannot interact, that there is only one bending mode, and that this mode will be both infrared and Raman active. These results also apply rigorously to real systems $L_x MCO$, provided that the atoms M, C, O are colinear and lie on a 3-fold or higher rotational axis.

For the bending mode, Eq. (2.35) becomes

$$F_\beta \left[m_M^{-1} r_{MC}^{-2} + m_O^{-1} r_{CO}^{-2} + m_C^{-1} (r_{MC}^{-1} + r_{CO}^{-1})^2 \right] - 5·8915 \times 10^{-5} v_\beta^2 = 0.$$

$$(3.1)$$

(For the derivation of the G element, see [13], Section 4.2.) Substitution of the chosen values into Eq. (3.1) gives bending frequencies

$$v_\beta \, (^{60}M^{12}C^{16}O) = 402·8 \, \text{cm}^{-1}$$
$$v_\beta \, (^{60}M^{13}C^{16}O) = 390·0 \, \text{cm}^{-1}$$
$$v_\beta \, (^{60}M^{12}C^{18}O) = 397·9 \, \text{cm}^{-1}$$

$$(3.2)$$

$$v_\beta \, (^{60}M^{13}C^{16}O) \, / \, v_\beta \, (^{60}M^{12}C^{16}O) = 0·9704$$
$$v_\beta \, (^{60}M^{12}C^{18}O) \, / \, v_\beta \, (^{60}M^{12}C^{16}O) = 0·9877.$$

$$(3.3)$$

The frequencies of Eq. (3.2) obviously lack any general application, and inspection of Eq. (3.1) shows that the results of Eq. (3.3) are dependent both on the mass of M and on the ratio of the MC and CO distances. Despite this, the values quoted may be taken as typical of a pure bending mode.

MC and CO Stretching Modes

For the stretching modes of an MCO group, Eq. (2.35) takes the form

$$\begin{bmatrix} F_{MC} & F_{MC,CO} \\ F_{MC,CO} & F_{CO} \end{bmatrix} \begin{bmatrix} m_M^{-1} + m_C^{-1} & -m_C^{-1} \\ -m_C^{-1} & m_O^{-1} + m_C^{-1} \end{bmatrix} - \begin{bmatrix} \lambda & 0 \\ 0 & \lambda \end{bmatrix} = 0. \quad (3.4)$$

Here the **F**-matrix is fixed by definition, while the **G** matrix is derived as in [13, Section 4.2]. For the particular atomic masses and chosen force constants, this equation leads to frequencies

$$v_1 = 2042 \cdot 7 \, \text{cm}^{-1}$$
$$v_2 = 420 \cdot 4 \, \text{cm}^{-1}. \quad (3.5)$$

The forms of the normal vibrations are

$$\delta (MC)_1 : \delta (CO)_1 :: -0 \cdot 580 : 1$$
$$\delta (MC)_2 : \delta (CO)_2 :: 1 : 0 \cdot 028 \quad (3.6)$$

(where $\delta(AB)$ is the relative change in the AB bondlength), and the potential energy distributions are given by

$$v_1(MC) : v_1(CO) : v_1(MC, CO) :: 3 \cdot 9 : 98 \cdot 5 : -2 \cdot 4$$
$$v_2(MC) : v_2(CO) : v_2(MC, CO) :: 98 \cdot 4 : 0 \cdot 6 : 1 \cdot 0. \quad (3.7)$$

The Energy Factoring Approximation

The results for the model MCO system given in Eqs (3.4) to (3.7) provide a test of the energy factoring (group frequency) approximation and an insight into its workings.

If coupling between the metal–carbon and carbon–oxygen stretching vibrations is assumed to have no effect on frequencies, then the values of Eq. (3.5) can be inserted directly into Eq. (2.9) to give *stretching parameters*

$$k(MC) = 198 \, \text{Nm}$$
$$k(CO) = 1685 \cdot 0 \, \text{Nm}. \quad (3.8)$$

(These parameters are calculated taking $1/\mu(CO)$ as $1/m_C + 1/m_O$, and $1/\mu(MC)$ as $1/m_M + 1/(m_C + m_O)$.)

Metal–carbon stretching parameters are not often quoted, since the relevant bands are sometimes difficult to assign, and are always, in real systems, affected by coupling with bending modes (see Section 3.3 below).

Carbon–oxygen stretching parameters, however, are used extensively to measure electron availability at the metal. If this is high, the metal will be a poor σ-acceptor and a good π-donor, and F_{CO} is therefore expected to be low. It is commonly assumed that trends in the stretching parameter faithfully reflect such trends in the force constant. This assumption is difficult to prove, but is powerfully supported, both by the numerical closeness between $k(CO)$ and F_{CO} in our model, and by the following, more detailed, analysis.

Equations (3.6, 3.7) show why the agreement between $k(CO)$ and F_{CO} is so good. Exact energy factoring would imply that unit expansion of the CO band should impose a constriction on the more deformable MC bond of 16/28 (0·571) units, so that the baricentre of the CO group does not move. This kinematic coupling produces the off-diagonal elements of the **G**-matrix in Eq. (3.4). Extension of the MC bond should produce no change in the C—O bondlength, since the carbonyl group would be acting as a rigid unit. In real systems, since the metal–carbon bond is not infinitely deformable, contraction imposed on the metal–carbon bond by an extension of the vibrating CO group does increase the potential energy of the system. This effect, however, is largely offset by the electronic interaction of MC and CO bonds, giving rise to cross-terms in the **F**-matrix. Extension of the C—O bond makes the ligand a better π-acceptor, and thus reduces the equilibrium metal–carbon bondlength.

The energy factoring approximation assumes no coupling between M—C and C—O deformations. Such coupling does occur, through both **F** and **G** matrices, but the approximation still works well, as the above analysis shows, because the effects of the two kinds of coupling are opposed. The **G**-matrix coupling is the same for all MCO groups. The **F**-matrix coupling no doubt varies, but the variation is likely to be systematic. A low value of F_{CO} suggests extensive metal–carbon π-bonding, which is expected to increase $F_{MC, CO}$. Thus $k(CO)$ depends on F_{CO}, on $F_{MC,CO}$ which is subject to the same influences as F_{CO}, and on $G_{MC, CO}$, which is constant. The deviation between F_{CO} and $k(CO)$ will not vary arbitrarily from one compound to another, and the use of $k(CO)$ in place of F_{CO} in comparative studies appears to be permissible.

Isotopic Substitution

The effects of isotopic labelling on the MCO bending mode are given by Eq. (3.2). Isotopic substitution at carbon affects the frequency more than does substitution at oxygen, even though the relative change in mass is less. This is a purely geometric effect, the coefficient of m_C^{-1} in Eq. (3.1) being greater than that of m_O^{-1}. The amplitude of oscillation of carbon is far greater than that of oxygen, since the moments of oscillation of these two atoms round the far heavier atom M must approximately cancel.

For the stretching modes, the frequencies of isotopically labelled species may be found using Eq. (3.3). The results of this calculation are

$$^{60}M^{13}C^{16}O:- \qquad v_1 = 1996 \cdot 3 \qquad v_2 = 415 \cdot 6$$

$$^{60}M^{12}C^{18}O:- \qquad v_1 = 1995 \cdot 2 \qquad v_2 = 410 \cdot 4$$

$$v_1\ (^{60}M^{12}C^{16}O) : v_1\ (^{60}M^{13}C^{16}O) : v_1\ (^{60}M^{12}C^{18}O) :: 1.0 \cdot 9773 : 0 \cdot 9767$$

$$v_2\ (^{60}M^{12}C^{16}O) : v_2\ (^{60}M^{13}C^{16}O) : v_2\ (^{60}M^{12}C^{18}O) :: 1.0 \cdot 9887 : 0 \cdot 9762.$$

$$(3.9)$$

If the energy factoring approximation were exact, then the ratios of frequencies would be

$$1 : 0 \cdot 9777 : 0 \cdot 9759 \quad (v_1) \qquad\qquad (3.9a)$$

and

$$1 : 0 \cdot 9882 : 0 \cdot 9770 \quad (v_2). \qquad\qquad (3.9b)$$

Several features of interest become apparent on comparing the results of (3.2), (3.8), and (3.9). The bending and M—C stretching modes occur at similar frequencies; they can in principle be distinguished by the *far higher* sensitivity of the former to isotopic substitution at carbon. The CO stretching modes of $M^{13}CO$ and $MC^{18}O$ occur extremely close together, the observed separation between them being only $1 \cdot 2$ cm^{-1}, or about a third of that required by the energy factored approximation. This has been shown [24] to be a consequence of the coupling of MC and CO stretching modes. Replacement of ^{16}O by ^{18}O slightly increases coupling by reducing the difference in frequency between the pure modes; thus $v(C^{18}O)$ is slightly higher, and $v(M—C^{18}O)$ slightly lower, than would have been predicted. Similar results would follow from the replacement of ^{12}C by ^{13}C; but these are slightly more than counterbalanced by the *reduction* in the magnitude of coupling through the off-diagonal term of the **G**-matrix.

Determination of Force Constants from Frequencies

The observed frequencies of $M^{12}C^{16}O$ cannot suffice to fix values for the three force constants. It is, however, possible to express any two of the force constants in terms of the third, and to calculate corresponding frequencies for the ^{13}C-substituted species.

The results of this exercise are given in Fig. 3.1, $F_{MC, CO}$ being chosen as the independent variable. The calculated value of F_{MC} is found to be relatively insensitive to the assumed value of $F_{MC, CO}$, but F_{CO} is much more affected. This is in accord with the degree of mixing. The lower frequency mode is close to being a pure metal–carbon stretching vibration, and is not greatly affected by the degree of coupling. A more disappointing, if no more surprising result is the insensitivity of the frequencies of the isotopically substituted species. Thus a change in $F_{MC,CO}$ from 50 to 100 Nm^{-1}, imposing a change of 59 Nm^{-1} in F_{CO}, shifts the M—^{13}C and ^{13}C—O frequencies by 0·26 and 1·24 cm respectively. Such a shift in ν(M—^{13}C) is barely detectable, while very small errors in ν(^{13}C—O) (or in ν(^{12}C—O)) will lead to errors in the calculated value of the CO force constant as large as the deviation between this constant and the parameter k(CO). This result was only to be expected; the changes in F_{MC} and F_{CO} are constrained to exactly counterbalance the effects on ^{60}M^{12}C^{16}O of variation in $F_{MC,CO}$, and this balance is almost as effective for the very similar species ^{60}M^{13}C^{16}O.

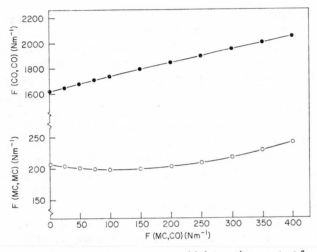

FIG. 3.1. Variation of calculated force constants with interaction constant for frequencies of Eq. (3.5).

Substitution with C^{18}O is slightly more helpful. A change of 5 Nm^{-1} in $F_{MC,CO}$, which requires F_{CO} to alter by 59 Nm^{-1}, raises the CO stretching frequency of ^{60}M^{13}C^{16}O by 0·12 cm^{-1}, and lowers the corresponding frequency of ^{60}M^{12}C^{18}O by 0·18 cm^{-1}. The effects of an assumed change in $F_{MC,CO}$ depend on the *relative* masses of carbon and oxygen, so that comparison of ^{13}CO- with C^{18}O-labelled molecules will lead to a better estimate of the force field than will comparison of either with the naturally occurring species. But even in this, the most favourable case, great accuracy of frequency measurement is necessary before the stretching constants are an improvement on the energy factored parameters. A good fit between observed and calculated frequencies is not sufficient evidence for the validity of a particular force field for the MCO unit. It is also necessary to determine the uncertainty of the force field derived, by using rules for the propagation of errors or by exploring effects of barely measurable shifts in frequency. This uncertainty may be disappointingly great. The same is obviously true for attempts to derive true force constants, rather than energy factored parameters, for species containing more than one CO group. The differences between the observed shifts for real molecules and the calculated shifts of the energy factoring approximation are small, but it is on the measurement of these small differences that any attempt to progress beyond the energy factoring approximation must depend.

3.3. Stretching Modes of *Trans* $M(CO)_2$

In the preceding section, the model system MCO was examined in some detail. This examination showed the effects of coupling between the MC and CO stretching modes, the relationship between the force constant for CO and the CO "energy factored" stretching parameter, the failure of the energy factored force field to give precise predictions of isotope effects, and the difficulty of uniquely defining the force field from frequency data. These considerations apply to all carbonyl complexes, but the presence of more than one CO group raises yet further questions. Thus, there will be more than one type of CO stretching mode, and it is not obvious whether these all interact with the MC modes to the same extent. The existence of interaction constants will ensure that, even if all the carbonyl groups are equivalent, the different modes will occur at different frequencies. This behaviour is represented in the energy factored force field by an interaction parameter, and it is of some interest to know how this is related to the interaction constants of the general valence force field. Finally, the effects of isotopic substitution on the different modes are not obvious; and these effects can be important in the study of systems where bands are infrared forbidden, especially if Raman data are not available. These considerations are illustrated here by the stretching modes of a linear model species $M(CO)_2$.

MC and CO Stretching Modes

As before, we shall assume values of 1700 Nm^{-1}, 200 Nm^{-1}, 70 Nm^{-1} and 60 for F_{CO}, F_{MC}, $F_{MC,CO}$ and the mass number of M. In addition, we shall assume values of 8, 50, and -10 Nm for $F_{CO,C'O'}$, $F_{MC,MC'}$ and $F_{MC,C'O'}$; these values are fairly close to those for the *trans* interaction in carbon tetrachloride solutions of $Cr(CO)_6$ [23].

The bending modes of *trans* $M(CO)_2$ belong to doubly degenerate representations of the point-group $D_{\infty h}$; they are thus precluded from interacting with the stretching modes, and will not be discussed further here. The stretching modes span the irreducible representations $\sum_g^+$ and $\sum_u^+$, corresponding to in-phase and out-of-phases combinations of the motions of the two carbonyl groups. The **F** and **G** matrices for these vibrations are given in Appendix II.

The chosen values for the force constants lead to the stretching frequencies

$$v_1(\Sigma_g^+, v(CO)) = 2064 \cdot 6 \text{ cm}^{-1}$$

$$v_2(\Sigma_g^+, v(MC)) = 386 \cdot 1 \text{ cm}^{-1}$$

$$v_3(\Sigma_u^+, v(CO)) = 2020 \cdot 7 \text{ cm}^{-1}$$

$$v_4(\Sigma_u^+, v(MC)) = 419 \cdot 3 \text{ cm}^{-1}.$$

(3.10)

The forms of the normal coordinates are

$$\delta(MC)_1 : \delta(CO)_1 :: -0 \cdot 584 : 1$$

$$\delta(MC)_2 : \delta(CO)_2 :: \quad 1 : 0 \cdot 051$$

$$\delta(MC)_3 : \delta(CO)_3 :: -0 \cdot 573 : 1$$

$$\delta(MC)_4 : \delta(CO)_4 :: \quad 1 : 0 \cdot 004.$$

(3.11)

Here $\delta(MC)$ and $\delta(CO)$ are the changes in metal–carbon and carbon–oxygen distance within a single MCO unit. The changes in the other unit are identical for vibrations 1 and 2, and opposed for vibrations 2 and 3. The potential energy distributions, summing over both the CO groups and their interactions are given by

$$v_1{}^{MC} \quad : v_1{}^{CO} \quad : v_1{}^{MC,CO} : v_1{}^{MC,MC'} : v_1{}^{MC,C'O'} : v_1{}^{CO,C'O'} ::$$

$$3 \cdot 86 \quad : 96 \cdot 70 \quad : -2 \cdot 32 : 0 \cdot 96 \quad : 0 \cdot 33 \quad : 0 \cdot 46$$

$$v_2{}^{MC} \quad : v_2{}^{CO} \quad : v_2{}^{MC,CO} : v_2{}^{MC,MC'} : v_2{}^{MC,C'O'} : v_2{}^{CO,C'O'} ::$$

$$77 \cdot 64 \quad : 1 \cdot 74 \quad : 1 \cdot 39 \quad : 19 \cdot 41 \quad : -0 \cdot 20 \quad : 0 \cdot 01$$

$$v_3{}^{MC} \quad : v_3{}^{CO} \quad : v_3{}^{MC,CO} : v_3{}^{MC,MC'} : v_3{}^{MC,C'O'} : v_3{}^{CO,C'O'} ::$$

$$3 \cdot 88 \quad : 100 \cdot 27 : -2 \cdot 37 : -0 \cdot 97 \quad : -0 \cdot 34 \quad : -0 \cdot 47$$

$$v_4{}^{MC} \quad : v_4{}^{CO} \quad : v_4{}^{MC,CO} : v_4{}^{MC,MC'} : v_4{}^{MC,C'O'} : v_4{}^{CO,C'O'} ::$$

$$133 \cdot 06 : 0 \cdot 01 \quad : 0 \cdot 17 \quad : -33 \cdot 26 \quad : 0 \cdot 02 \quad : -0 \cdot 00$$

(3.12)

Despite the positive sign of the interaction constant $F_{MC,MC'}$, the out-of-phase metal–carbon vibration occurs at higher frequency than does the corresponding in-phase vibration. This is a consequence of the difference between G_{33} and G_{44}, *i.e.* of the finite mass of the metal and the resultant kinematic coupling of distortions of the two separate metal–carbon bonds; and the ordering

$$v \text{ (antisymmetric)} > v \text{ (symmetric)}$$

is usual for metal–ligand vibrations [25]. In real systems $trans$-M(CO)$_2$L$_4$, the antisymmetric metal–carbon stretch is kinematically coupled to angular distortions involving the other ligands, and the evaluation of metal–carbon stretching parameters is therefore inappropriate.

The above considerations apply only minimally to the CO modes. For these it proves useful to evaluate energy factored stretching and interaction parameters, defined as the solutions to a model, energy-factored, problem using the actual frequencies. Such a model is greatly simplified by the lack of direct kinematic coupling between the different CO groups, so that the **G**-matrix is diagonal:

$$\mathbf{G}(EF) = [(1/m_c + 1/m_0)_i \delta_{ij}]. \tag{3.13}$$

For all $-\ {}^{12}C^{16}O$ molecules, (3.13) may be combined with the energy-factored form of (2.34) to give a secular determinant

$$|k_{ij} - 0 \cdot 40383 \times 10^{-4} v^2 \delta_{ij}| = 0 \tag{3.14}$$

where k_{ij} is the element of the energy factored force field connecting coordinates i and j.

For the species M(CO)$_2$, the energy-factored **F**-matrix may readily be expressed in terms of symmetry coordinates:

$$\mathbf{F}(EF) \text{ (symm)} = k(CO) + k(CO, C'O')$$
$$\mathbf{F}(EF) \text{ (antisymm)} = k(CO) - k(CO, C'O'). \tag{3.15}$$

Comparing Eqs (3.14, 3.15) with the frequencies of Eq. (3.10) gives

$$k(CO) = 1685 \cdot 2 \ \mathrm{Nm}^{-1}$$
$$k(CO, C'O') = 36 \cdot 2 \ \mathrm{Nm}^{-1}. \tag{3.16}$$

This is an important result. It shows that the stretching parameter takes almost the same value for this system as it did for the model system MCO. It is not significantly perturbed by coupling between the carbonyl groups, and for polycarbonyls is related to the force constant in just the same way as for monocarbonyls.

The interaction parameter $k(CO, C'O')$ is also an important quantity. It originates in part from $F_{CO,C'O'}$, but mainly from the interaction constants $F_{MC,C'O'}$ and $F_{MC,MC'}$, through the metal–carbon bondlength imposed by the CO vibration. The changes in metal–carbon bondlength are determined by the relative masses of carbon and oxygen,† and are thus similar in all carbonyls, while the interaction constants involved are all valid measures of electronic coupling. Thus the interaction parameter is a valid measure of the overall coupling between different MCO units.

Equations (3.11, 3.12) may be compared with the corresponding Eqs (3.6) and (3.7), for a single MCO group. The degree of distortion of the metal–carbon bond for unit displacement of the CO group is almost the same in all cases, as is the potential energy distribution for such displacement within each MCO unit. These findings confirm that the stretching parameter has the same significance for $M(CO)_2$ as it has for MCO. It would thus appear that stretching parameters may be usefully compared, even between systems that differ in the number and spatial distribution of carbonyl groups.

Isotopic Substitution

There are six force constants for the stretching modes of linear $M(CO)_2$, and only four stretching frequencies, so that the force-field is under-determined. The situation is even worse for non-linear species, for which bend–stretch interactions must be considered. The force fields can in principle be found by isotopic substitution, but the results of Section 3.2 show that this approach can be invalidated by quite small errors in frequency shifts. It has been pointed out that such errors can arise, in unsymmetrically labelled species, as a result of anharmonicity [26]. The present state of our knowledge about the anharmonicity of metal carbonyls is not, in the author's view, satisfactory, and the topic is discussed at greater length in Chapter 6. For the moment, it is important to note only that anharmonicity causes observed frequencies to be slightly lower than the mechanical frequencies of an exact quadratic force field, and that the deviations are related both to the amplitudes of individual vibrations and to cross-terms. The forms of the vibrations of $M(^{12}C^{16}O)_2$, $M(^{13}C^{16}O)_2$ and $M(^{12}C^{18}O)_2$ are very similar, consisting of symmetric and antisymmetric combinations of the motions of the individual MCO groupings. The normal modes of asymmetrically substituted species will be rather different, since the two differently weighted MCO units will not contribute equally. This could lead to differences in anharmonicity errors, which would vitiate the results

---

† It follows that the CO stretching parameter should be to some small extent sensitive to symmetrical isotopic substitution. $^{13}C^{16}O$-containing species should show slightly lower interaction parameters than the $^{12}C^{16}O$ analogs, while those for $^{12}C^{18}O$-containing species should be higher. The actual frequencies for the metal hexacarbonyls [23] may be used to calculate the relevant parameters, which bear out these expectations; but the effects on the interaction parameters are at most a few per cent.

of any full analysis. Thus, exact studies should be confined to the three symmetrically weighted species $M(^{12}C^{16}O)_2$, $M(^{13}C^{16}O)_2$ and $M(^{12}C^{18}O)_2$ ($^{13}C^{18}O$ is not readily available). The problem may then be solved, using the **F** and **G** matrices of Appendix II, by finding the set of force constants that give the best fit to the frequencies. This task is facilitated by the relative insensitivity (see Section 3.2 above) of the metal–carbon stretching frequencies to the size of the interaction constants linking metal–carbon and carbon–oxygen modes.

Symmetric isotopic substitution is of no value in the determination of energy factored parameters, since within the energy factoring approximation its only effect is to shift all the CO stretching frequencies by an equal and predictable factor. Asymmetric substitution can, however, be very useful in such investigations, especially where some of the vibrations are infrared inactive. For example, in linear $M(CO)_2$, v_1 is infrared inactive. It is, of course, Raman active, but the compound may be coloured or photosensitive, the Raman band may be weak (as for v_1 in $V(CO)_6^-$ [27]), or instrumentation may not be available. The species $M(^{12}C^{16}O)$ ($^{13}C^{16}O$) belongs to the point-group $C_{\infty v}$ and both the CO stretching modes are allowed. The higher frequency mode corresponds to an in-phase motion of the ^{12}CO and ^{13}CO groups, and is therefore moderately weak in the infrared spectrum; it will, however, be readily observable in isotopically enriched material. The lower frequency mode is predominantly an out-of-phase combination, and is generally observable, even at natural (1·1%) abundance of C, as a low-frequency "satellite" band.

The analysis of the frequency data may then be carried out in one of several ways. For the labelled species, the secular equation may be written as

$$\left| \begin{bmatrix} k & i \\ i & k \end{bmatrix} \begin{bmatrix} 1/\mu & 0 \\ 0 & 1/\mu^* \end{bmatrix} - \lambda\,\mathbf{E} \right| = 0 \qquad (3.17)$$

where μ and μ^* are the effective masses [Eq. (2.3)] of normal and labelled carbonyl.

The **FG** matrix of Eq. (3.17) is non-Hermitian, *i.e.* $(\mathbf{FG})_{ij}$ is not identical with $(\mathbf{FG})_{ji}$. The equation may be put into a more elegant form [28] by a similarity transform. Multiplying both **FG** and $\lambda\,\mathbf{E}$ to the left and to the right by $G^{\frac{1}{2}}$ and $G^{-\frac{1}{2}}$ respectively gives

$$\left| \begin{bmatrix} \mu^{-\frac{1}{2}} & 0 \\ 0 & (\mu^*)^{-\frac{1}{2}} \end{bmatrix} \begin{bmatrix} k\mu^{-1} & i(\mu^*)^{-1} \\ i\mu^{-1} & k(\mu^*)^{-1} \end{bmatrix} \begin{bmatrix} \mu^{\frac{1}{2}} & 0 \\ 0 & (\mu^*)^{\frac{1}{2}} \end{bmatrix} - \lambda\,\mathbf{E} \right| = 0, \qquad (3.18)$$

(since $G^{\frac{1}{2}} EG^{-\frac{1}{2}} = E$), which becomes

$$\begin{vmatrix} k\mu^{-1} & i(\mu\mu^*)^{-\frac{1}{2}} \\ i(\mu\mu^*)^{-\frac{1}{2}} & k(\mu^*)^{-1} \end{vmatrix} - \lambda E = 0. \tag{3.19}$$

This procedure is quite general. Thus it is always possible to write the **FG** matrix for an all $-{}^{12}C^{16}O$ species, and convert this to the $G^{\frac{1}{2}}FG^{\frac{1}{2}}$ matrix for a labelled species by multiplying every row and every column involving the labelled species by a factor of $\mu^{-\frac{1}{2}}(\mu^*)^{\frac{1}{2}}$. The procedure may also be applied to a matrix written in terms of the symmetry coordinates of the labelled species. Thus for a species $M(CO)_5 (CO)^*$, it is possible to form symmetry coordinates belonging to the representations E and B_2 of C_{4v}, correlating with those of symmetry T_1 and E in $M(CO)_6$, and unaffected by the isotopic labelling. The effects of this labelling are confined to modes of symmetry A_1 in C_{4v}, and can be analysed using as symmetry coordinates the displacements of $(CO)^*$ and of CO *trans* to $(CO)^*$, and the in-phase combination of displacements of the four CO groups *cis* to $(CO)^*$.

In principle, it is possible to express i in terms of k and the infrared frequency of the unlabelled species; then (3.19) contains only one unknown parameter and can be solved by equating $0.58915 \times 10^{-4}v$ (satellite) with the lower frequency root. This method of determining the force field is studied in Section 3.6 below, with special reference to species of the type $M(CO)_2$; it is shown there that it is not acceptable, since small errors in frequency measurement are magnified enormously in the calculated force field. It is better to use both frequencies for the isotopically labelled species and Eq. (3.19). Better still, perhaps, is to use the frequencies for the labelled species in conjunction with v_2 for the unlabelled species to find v_1. The *Teller–Redlich product rule* gives a convenient method of doing this.

The product rule is based on the properties of determinants and generally involves all the frequencies, including notional frequencies corresponding to rotation and translation of the entire molecule in a vanishingly weak lattice field. Such notional frequencies are not involved in the energy-factored force field, and the rule takes the simple form

$$\Pi(\omega^*/\omega) = \Pi(\mu/\mu^*)^{n/2} \tag{3.20}$$

The product runs over each individual frequency, degenerate frequencies being repeated the appropriate number of times. μ and μ^* are the reduced masses of the normal and n labelled CO groups. The ω are mechanical frequencies corrected for anharmonicity; the error introduced by the use of the observed frequencies v is, however, probably small. The "theoretical values" for $(\mu/\mu^*)^{\frac{1}{2}}$ are 0.9777 and 0.9759 for ${}^{13}CO$ and $C^{18}O$ respectively compared with ${}^{12}C^{16}O$, but the results of isotopic substitution on a realistic mode system MCO shown that "practical ratios" are preferable, and experiment confirms this prediction. The force field of Eq. (3.8) implies practical ratios

$$R(^{13}CO) = [\mu(^{12}C^{16}O)/\mu(^{13}C^{16}O)]^{\frac{1}{2}}_{EF} = 0.9773$$
$$R(C^{18}O) = [\mu(^{12}C^{16}O)/\mu(^{12}C^{18}O)]^{\frac{1}{2}}_{EF} = 0.9767$$

(3.21)

Some observed values of $R(^{13}CO)$ are 0.9774 [24] in $CoCl_2$ $(PEt_3)_2$ CO and 0.9766 [29] in $Mo(CO)_6$. Values of $R(C^{18}O)$ included 0.9776 in $CoCl_2$ $(PEt_3)_2$ CO, and 0.9770 and 0.9771 in *mono*-(axial) and *di*-(axial) substituted *cis* $Fe(CO)_4I_2$ [30]. Thus it appears that the practical values of Eq. (3.21) are preferable to those calculated *a priori* with complete neglect of MC–CO coupling, and may, if anything, slightly underestimate the coupling correction.

3.4. 'Exact' and 'Approximate' Energy Factored Force Fields

Exact Parameters

Metal carbonyls may be divided into two classes; those in which all the CO groups are physically equivalent, and those in which CO groups may be found in more than one kind of environment. The model systems of Sections 3.2 and 3.3 belong to the first class, as do systems $M(CO)_3$ (D_{3h} or C_{3v}), $M(CO)_4$ (tetrahedral, square planar, or distorted square planar [D_{2d}]) and $M(CO)_6$ (O_h, D_{3h} and D_{3d}). Systems $M(CO)_5$ (D_{3h} or C_{4v}), $M(CO)_4$ (C_{2v}, as in *cis*-$L_2M(CO)_4$), and $M(CO)_3$ (C_{2v}, as in *mer*-$L_3M(CO)_3$) belong to the second class.

In all these cases, an energy factored force field may be specified by relatively few parameters. For systems of the first class, there is only one stretching parameter. This is equal to the mean stretching parameter (defined below) which can be found if all the CO stretching frequencies are known. It is also possible to calculate all the interaction parameters for mononuclear molecules of this class, although this is sometimes not so for polynuclear species [31]. For systems of the second class there are two or more different stretching parameters, one for each chemically non-equivalent CO group. The number of interaction parameters is also higher than for species of the first class. It follows that, for species of the second class, the energy factored force field cannot be exactly evaluated except by partial isotopic substitution.

Mean Stretching Parameters

For all isotopically unsubstituted species, whether they belong to the first class or the second, the mean stretching parameter is given (in Nm^{-1}) by

$$\bar{k}(CO) = 4.0383 \times 10^{-4} \sum_i g_i v_i^2 / \sum_i g_i$$

(3.22)

Here g_i is the degeneracy of the ith CO stretching mode, of frequency v_i cm^{-1}; thus $\sum_i g_i$ is equal to the number of carbonyl groups. Equation (3.22) follows immediately from the invariance of the sum of the diagonal elements of a matrix (the character) to a similarity transformation, since one such transformation is the diagonalisation of an energy-factored F-matrix written in terms of the individual CO groups. To use Eq. (3.22), it is not necessary to know the exact assignment of all CO stretching frequencies but merely their degeneracy, and there are cases (such as cis-$L_2M(CO)_4$) where the former may be in doubt although the latter is not. For species of the first class, the frequency given by Eq. (3.22) is obviously equal to the unique stretching parameter, and cases can arise [31, 32] where this parameter can be found even though the full energy factored force field is underdetermined.

Cotton–Kraihanzel and Other Approximate Force Fields

Mutually *trans* carbonyl ligands in an octahedral complex share two metal d-orbitals, while groups mutually *cis* share only one. This fact led many years ago to the suggestion [5] that the interaction in the former case might be close to twice that in the latter. While this is now known to be very far from true for interaction *constants* [33], the suggestion has been extended with considerable success to the interaction *parameters* of substituted species [34, 35]. For such species it is further argued that the stretching parameter of a CO group *trans* to another CO group should be higher than that of a CO group *trans* to almost any other ligand, and that the various interaction parameters are expressible in terms of a single parameter. In the notation of Section 3.6

$$k(2) > k(1) \tag{3.23}$$

$$c \sim d = i \tag{3.24}$$

$$t = 2c. \tag{3.25}$$

Parameters evaluated using Eq. (3.24, 3.25) are known as *Cotton–Kraihanzel (CK) parameters*.

An alternative set of approximations has been proposed by some authors [36, 37]; these fail, however, to satisfy the condition that for symmetric carbonyls c is identical with d, and must therefore be rejected. The failure to equate c and d in symmetric carbonyls may be due to an over-ambitious attempt to allow for the effects on bonding changes of relative angles, and a simpler and more satisfactory set of approximations has now been developed, combining the virtues of the Cotton–Kraihanzel and alternative earlier sets (see Appendix IV).

The 'Independent Variable' Method

In many carbonyls of the second class, one representation of the molecular point-group is spanned twice while the rest are spanned only once, with the result that the number of parameters to be determined exceeds the number of frequencies by one. In these cases, it is possible to choose one independent variable in terms of which all the other parameters may conveniently be expressed. The secular equation for the doubly spanned representation can be written in the form

$$\begin{vmatrix} k(m) - K & k(\text{int}) \\ k(\text{int}) & k(n) - K \end{vmatrix} = 0 \tag{3.26}$$

where K is an effective parameter corresponding to the vibrational frequency v:

$$K = 4 \cdot 0383 \times 10^{-4} \, v^2. \tag{3.27}$$

Expanding Eq. (3.26) and solving for K in terms of $k(m) + k(n)$, $k(\text{int})$ gives

$$2K = k(m) + k(n) \pm \{[k(m) - k(n)]^2 + 4 k(\text{int})^2\}^{\frac{1}{2}}. \tag{3.28}$$

Let $k(m)$ be chosen as the independent parameter. If $K(1)$ $K(2)$ are the roots of (3.26), then

$$K(1) + K(2) = k(m) + k(n) \tag{3.29}$$

$$[K(1) - K(2)]^2 = [k(m) - k(n)]^2 + [2 k(\text{int})]^2$$

$$= [2 k(m) - K(1) - K(2)]^2 + [2 k(\text{int})]^2. \tag{3.30}$$

Since $K(1)$, $K(2)$ are known from the observed frequencies, Eq. (3.29) may be regarded as a linear equation for $k(n)$ in terms of $k(m)$. Eq. (3.30) connects $k(m)$ and $k(\text{int})$; the plot of $k(\text{int})$ against $k(m)$ is a circle of radius $[K(1) - K(2)]/2$, centred at the point $\{k(m) = [K(1) + K(2)]/2; k(\text{int}) = 0\}$. It is obvious that real roots may be found for $k(\text{int})$ only if $k(m)$ and $k(n)$ lie between the limits $K(1)$ and $K(2)$, and that $k(\text{int})$ takes a maximum value of $[K(1) - K(2)]/2$. Although Eq. (3.30) may be formally satisfied by negative as well as by positive values of $k(\text{int})$, only positive values are physically acceptable for interaction parameters linking CO groups on the same atom.

Since π-bonding is often taken to increase interaction constants, the particular solution

$$\{k(m) = k(n) = [K(1) + K(2)]/2; k(\text{int}) = [K(1) - K(2)]/2\}$$

may be taken to represent the greatest degree of π-bonding consistent with the observed frequencies, and has been commended on these grounds. The author has misgivings about such a procedure, since the observed frequencies are merely consequences of the bonding in the molecule and not constraints upon it. The independent parameter method is not useful primarily as a procedure for finding extremum values, but rather because it exposes the consequences of *any* simplifying procedure, such as the adoption of either of the CK relationships. In particular, it is sometimes found that Eqs (3.24) and (3.25) are not satisfied within that range of values of the independent variable that leads to real roots. In other words, *there are cases where one or both of the Cotton–Kraihanzel relationships leads to unreal roots.*

Where there are three physically distinct interaction constants, as in systems $LM(CO)_5$, *cis*-$L_2M(CO)_4$, the CK relationships impose two constraints on the system. But it may be that (as in the examples cited) knowledge of all the vibrational frequencies leaves only one degree of freedom. In such cases *it is possible to satisfy analytically the frequency data and one or other of the CK relationships, but not both.*

These difficulties in the application of the CK relationships may be met by suppressing some of the frequency data, by finding the CK parameters that give a "best" (e.g. least mean square deviation) fit to all the data, or by dropping one of the relationships. The author feels strongly opposed to the first of these procedures and to some extent to the second, since the parameters found will not exactly obey Eq. (3.22). The best procedure in these cases is probably to discard condition (3.24), which is known [38] to lead to unreal roots for a number of molecules.

3.5. Relative Infrared Intensities in the Energy Factored Approximation

The integrated intensity I_k of a carbonyl infrared absorption band is given by

$$I_k = (N\pi/3c^2)g_k \, (\partial\mu/\partial Q_k)^2 \tag{3.31}$$

if we ignore the small effects of anharmonicity. Here N is Avogadro's number and c the speed of light, g_k is the degeneracy of the vibration, and $\partial\mu/\partial Q_k$ is the rate of change of molecular dipole moment with a mass-weighted normal coordinate associated with the vibration.

For Eq. (3.31) to be useful, the quantity $\partial\mu/\partial Q_k$ must be re-expressed in terms of the individual local oscillating dipoles of each MCO group. Using Eq. (2.41) the identity

$$\partial\mu/\partial Q_k = \sum_i (\partial\mu/\partial S_i) \, (\partial S_i/\partial Q_k) \tag{3.32}$$

becomes

$$\partial\mu/\partial Q_k = \sum_j L_{jk} \, (\partial\mu/\partial S_1). \tag{3.33}$$

Unit distortions of the symmetry coordinates may be expressed in terms of unit distortions of the individual CO groups.

$$\Delta(S_j) = \Sigma l_{ij} \, \Delta(r_i) \tag{3.34}$$

where coefficients l_{ij} are determined by symmetry considerations only. Then the individual bond distortions are given by

$$r_i = \Sigma l_{ij} \, S_j \tag{3.35}$$

and Eq. (3.33) becomes

$$\partial\mu/\partial Q_k = \sum_i \sum_j l_{ij} \, L_{jk} \, \partial\mu/\partial r_i \tag{3.36}$$

It is important to note that the summations in Eq. (3.36) are vector additions, and that the coefficients in Eq. (3.36) must be normalised, i.e. multiplied by scaling factors so that in all cases

$$\sum_i (l_{ij})^2 = 1$$

$$\sum_j (L_{jk})^2 = \mu^{-1} \tag{3.37}$$

Where only one kind of CO group is present, Eqs (3.31) and (3.36) make it possible to express the relative intensities of all the bands in terms of geometric factors alone; thus angles between the directions of the local oscillating dipoles may be found. These are not necessarily the same as bond angles, since the local oscillating dipole of a bond that does not lie on a rotational axis need not be colinear with that bond [39]. When more than one kind of CO group is present, it may be possible to infer from intensity and frequency data the relative magnitudes of the different oscillating dipoles, as well as the angles between them [40]. Such calculation requires all the parameters of the energy factored force field to be known, and thus presupposes a study using frequencies of isotopically labelled species. It must also be presumed that the energy factored force field gives a good description of the mixing of different coordinates of the same symmetry, despite the complex nature of the interaction parameters.

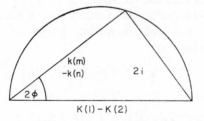

FIG. 3.2. The independent variable method of Eq. (3.38).

The mixing coefficients of two modes of the same symmetry can conveniently be found in a form suitable for the calculation of intensities by a form of the "independent variable" method described in Section 3.4. Equation (3.30) shows that $k(m) - k(n)$ and $2k(\text{int})$ may be represented by two sides of a right-angled triangle, of which the hypotenuse is $K(1) - K(2)$ (Fig. 3.2). The quantity $\cos 2\phi$ may be chosen as the independent variable, in terms of which the parameters may be expressed:

$$k(m) = [K(1)(1 + \cos 2\phi) + K(2)(1 - \cos 2\phi)]/2$$

$$k(n) = [K(1)(1 - \cos 2) + K(2)(1 + \cos 2\phi)]/2$$

$$k(\text{int}) = [K(1) - K(2)][1 - \cos^2 2\phi]^{\frac{1}{2}}/2$$

$$= [K(1) - K(2)]\sin 2\phi/2. \tag{3.38}$$

It can be shown that, if the symmetry coordinates S_1, S_2 are so chosen that $k(m) = F_{11}$, $k(n) = F_{22}$, then

$$L_{11} = L_{22} = \mu^{-\frac{1}{2}}\cos\phi$$

$$L_{21} = -L_{12} = \mu^{-\frac{1}{2}}\sin\phi. \tag{3.39}$$

Allowing 2ϕ to run from $0°$ to $180°$ (since $k(\text{int})$ is presumed positive) then gives all possible values of the mixing coefficients consistent with the frequencies of the isotopically unsubstituted molecule (Fig. 3.3). Equations

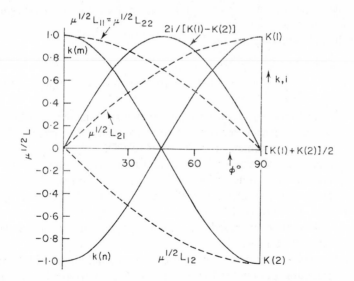

FIG. 3.3. Parameters and mixing coefficients as a function of the independent parameter.

(3.38, 3.39) are used extensively in Section 3.6 in discussions of infrared intensities for carbonyls of the second class.

Whether more than one type of CO group is present or not, Eq. (3.36) is equivalent to the first term of a Taylor expansion for the change in dipole, and implies that interaction terms are ignored. It has been shown that such interaction terms, if appreciable, would seriously affect the calculated angles [41]. Finally it must be emphasised that the theory given in this section refers to the *integrated intensities* of bands. The determination of such integrated intensities is not easy; the experimental problems are discussed in Chapter 4. It is not acceptable to use the maximum extinction coefficient as even the most qualitative measure of intensity, since (as discussed in Chapter 5) bands can differ considerably both in shape and width.

3.6. Force Fields, Parameters, Normal Coordinates and Relative Intensities for Species $L_xM(CO)_y$ in the Energy Factored Approximation

This Section contains analytical expressions for the energy factored force field in terms of the carbonyl frequencies, as well as for the forms of the normal coordinates and the relative infrared intensities, in the most commonly encountered cases. Selection rules for infrared and Raman spectra are also quoted, and the possible assignments of the observed frequencies are discussed. Where the problem is underdetermined, general parametric solutions are given and the consequences of simplifying approximations examined.

The following symbols are used:

μ Reduced mass of CO.

K Effective parameter, $4\cdot0383 \times 10^{-4} \nu^2$, corresponding to observed frequency.

$\bar{k}$ Mean effective parameter, equal to mean parameter. (See Eq. 3.22).

$k(1)$ Stretching parameter for CO *trans* to some other ligand.

$k(2)$ Stretching parameter for CO *trans* to CO.

t Interaction parameter for mutually *trans* CO groups.

c Interaction parameter for mutually *cis* CO groups.

d Interaction parameter for mutually *cis* CO groups, one only of which is *trans* to CO.

i Approximate averaged *cis* interaction parameter; unique interaction parameter.

$k(a)$ Stretching parameter for axial CO in trigonal bipyramidal species.

$k(e)$ Stretching parameter for equatorial CO in trigonal bipyramidal species.

e Interaction parameter between axial and equatorial CO groups in trigonal bipyramidal species.

f Interaction parameter between two equatorial CO groups in trigonal bipyramidal species.

$\Delta(n)$ Unit displacement of CO group of stretching parameter $k(n)$.

$\mu(n)$ Magnitude of local dipole set up by the displacement $\Delta(n)$. (this must not be confused with μ).

$I(\Gamma)$ Integrated intensity of infrared band due to vibration of symmetry Γ.

$\theta(m)$ Angle between local oscillating dipoles of CO groups connected by the interaction constant m.

$R(^{13}CO)$, $R(C^{18}O)$ Practical values of $[\mu(^{12}C^{16}O)/\mu(^{13}C^{16}O)]^{\frac{1}{2}}$, $[\mu(^{12}C^{16}O)/\mu(^{12}C^{18}O)]$.

$M(CO)_2$

Case (i): if the two carbonyl groups are equivalent, the analysis is completely straightforward. The symmetry coordinates, which are also normal coordinates, are

$$S(+) = [\Delta(1) + \Delta(2)]/\sqrt{2}$$
$$S(-) = [\Delta(1) + \Delta(2)]/\sqrt{2}. \tag{3.40}$$

$S(+)$ belongs to Σ_g^+, A_1, A', or A_1' in the point-groups $C_{\infty v}$, C_{2v}, C_s, or D_{3h}, while $(S-)$ belongs to Σ_u^+, B_1, A'' or A_2''. The secular equations are

$$K(+) = k + i$$
$$K(-) = k - i \tag{3.41}$$

so that

$$k = [K(+) + K(-)]/2$$
$$i = [K(+) - K(-)]/2. \tag{3.42}$$

The relative infrared intensities are given by

$$I(+)/I(-) = \tan^2 (\theta/2) \tag{3.43}$$

so that, for $\theta = 180$, the symmetric stretch is vanishingly weak. It is, however, formally allowed *unless* the complex belongs to a point-group that includes C_i, S_u, or D_u. The antisymmetric stretch is always infrared allowed and intense. In the Raman spectrum, the symmetric stretch is always allowed and polarised. Its intensity, however, may depend on the relative signs of the changes in longitudinal and transverse polarisability on distortion of an individual CO group [42]. The antisymmetric stretch is Raman-forbidden in groups containing C_i, D_n ($n \geqslant 3$), or S_n ($n \geqslant 6$); in other cases it is formally allowed, and depolarised.

The secular equation for the species with one isotopically labelled CO group is

$$\begin{vmatrix} k - K & Ri \\ Ri & R^2k - K \end{vmatrix} = 0. \tag{3.44}$$

Putting i equal to $k - K(-)$, and expanding the determinant, gives

$$k = [K^2 - R^2 K(-)^2]/[(1 + R^2) K - 2R^2 K(-)] \tag{3.45}$$

In principle, the force field can be found by putting K equal to $4{\cdot}0383 \times 10^{-4} v(\text{satellite})^2$. Closer inspection of Eq. (3.45) shows, however, that this will be of doubtful value unless frequency measurement is extremely accurate. The numerator and denominator on the right-hand side of Eq. (3.45) are both small, and errors in K and $K(-)$ can have dramatic effects. This is related to the fact that K must lie between the narrow limits of $K(-)R^2$ (for $i \ll k(1 - R^2)$) and $K(-) (R^2 + 1)/2$ (for $i \gg k(1 - R^2)$).

Case (ii): if the two carbonyl groups are not equivalent, the force field is underdetermined with one degree of freedom. The secular equation and the derivation of the range of solutions follow Eqs (3.26–3.30, 3.38, 3.39). A solution may be specified by estimating i from related symmetric species, or (subject to the reservations of the preceding paragraph) by matching data for isotopically labelled species. If the force constants of the two CO groups are appreciably different, then their local oscillating dipoles may also differ. This possibility vitiates any attempt to use relative intensity data to find the form of the normal modes and hence the force field.

$M(CO)_3$

Case (i): if all the CO groups are equivalent the molecule must belong to the point-group C_3, or to some higher point-group (commonly C_{3v}) that includes this. Using the notation of Fig. (3.4), and the labels for C_{3v} symmetry, the symmetry coordinates may be written as

$$S(A_1) = [\Delta(1) + \Delta(2) + \Delta(3)]/\sqrt{3}$$
$$S(E)_x = [\Delta(2) + \Delta(3) - 2\Delta(1)]/\sqrt{6}$$
$$S(E)_y = [\Delta(2) - \Delta(3)]/\sqrt{2}. \tag{3.46}$$

The secular equations are

$$K(A_1) = k + 2i$$
$$k(E) \quad = k - i \tag{3.47}$$

and simple trigonometry shows that the relative intensities of the two bands
are given by

$$I(A_1)/I(E) = \tan^2 \phi$$
$$= [3 \cot^2 \theta/2 - 1]/4. \qquad (3.48)$$

The intensity of the A_1 mode vanishes when the three oscillating dipoles are
coplanar ($\phi = 0$, $\theta = 120°$), and the A_1 band becomes formally infrared
forbidden when the complex belongs to the point-group D_{3h}. The E band is
always infrared allowed. Both bands are always Raman-allowed, the A_1
band being polarised.

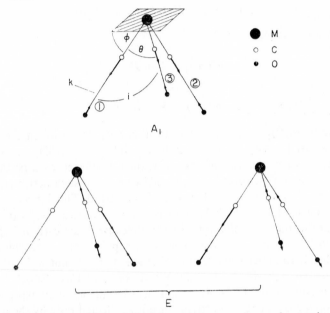

Fig. 3.4. Energy factored force field and symmetry coordinates for species M(CO)$_3$,
Case (i).

Case (ii): If one CO group is slightly different from the other two, then the
difference between the two physically distinct interaction constants can be
ignored. We may label the unique CO group as CO(1), and assign to it
stretching parameter $k + 2\delta$, while CO(2), CO(3) both have stretching
parameters $k - \delta$; δ may of course be either positive or negative. The secular
equations may then be written, using the labels for the point-group C_B, as

$$A': \begin{vmatrix} k + 2\delta - K & \sqrt{2}i \\ \sqrt{2}i & k - \delta + i - K \end{vmatrix} = 0. \qquad (3.49)$$

$$A'': K = k - i - \delta.$$

These equations may be rearranged to give

$$K(A')_1 + K(A')_2 + K(A'') = 3k$$
$$K(A')_1 + K(A')_2 = 2k + \delta + i$$
$$[K(A')_1 - k)] [K(A')_2 - k] = 2i\delta - 2\delta^2 - 2i^2. \tag{3.50}$$

In the limit of δ being very much smaller than i, terms in δ^2 may be ignored, so that expanding the quadratic of (3.50) gives

$$(K - k)^2 - (K - k)(\delta + i) - 2i^2 + 2i\delta - 2\delta^2 = 0$$
$$\doteqdot (K - k - 2i)(K - k + i - \delta)$$

and the roots become

$$K(A')_1 = k + 2i$$
$$K(A')_2 = k - i + \delta$$
$$K(A'') = k - i - \delta. \tag{3.51}$$

In this limit, $S(A')_1$ takes the same form as does $S(A_1)$ in Case (i), while $S(A_2)$ and $S(A'')$ correspond exactly to $S(E)_x$ and $S(E)_y$, respectively. The higher frequency mode of Case (i) is then unaffected by a perturbation that leaves the average force constant unchanged, but such perturbations do have the effect of splitting the lower frequency band. This limit of Case (ii) is useful as a model for the behaviour of symmetric tricarbonyls (Case i) in real solutions, since the solvation of each molecule of the carbonyl species is unlikely to be perfectly isotropic. The anisotropy of solvation, which varies from molecule to molecule, will not affect the A_1 mode but will cause slight splitting of the E mode. As a result, the A_1 band of Case (i) is far sharper than the E band. Indeed, the A_1 band is less broadened than expected for a band involving only one CO group, since it is affected only by the averaged values of the varying solvation at several different sites. This is a general phenomenon. Non-degenerate bands involving the motion of several carbonyl groups show less solvent broadening than do bands involving a single CO ligand, while degenerate bands are broadened most of all. These results are important in determining the characteristic appearance of spectra, in assignment, and in recognising and resolving overlapping patterns in the spectra of mixtures.

If δ is very much larger than i, the roots of the secular equation approach

$$K(A')_1 = k + 2\delta$$
$$K(A')_2 = k + i - \delta$$
$$K(A'') = k - i - \delta. \tag{3.52}$$

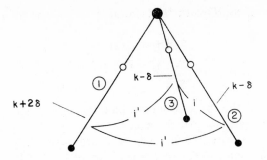

FIG. 3.5. Energy factored force field for $M(CO)_3$. Case (iii), [for case (ii), put $i' = i$].

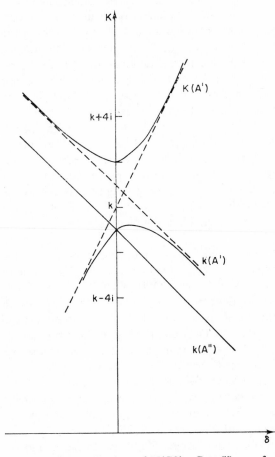

FIG. 3.6. Behaviour of the vibrations of $M(CO)_3$, Case (ii), as a function of δ.

and the normal coordinates tend towards the symmetry coordinates of Fig. 3.5:

$$S(A')_1 = \Delta(1)$$
$$S(A')_2 = [\Delta(2) + \Delta(3)]/\sqrt{2}$$
$$S(A'') = [\Delta(2) - \Delta(3)]/\sqrt{2}. \qquad (3.53)$$

The behaviour of the roots in intermediate cases is shown in Fig 3.6.

While the energy factored force field for Case (ii) is determinate, two kinds of ambiguity arise; the first of these concerns the assignment of bands, while the second is due to the quadratic nature of Eq. (3.50). Formally, there is no way of distinguishing between the order $K(A')_1 > K(A')_2 > K(A'')$ (with δ positive) and $K(A')_1 > K(A'') > K(A')_2$ (with δ negative). In practice, however, a choice can sometimes be made using model compounds, and assuming that the *trans*-influences of ligands on carbonyl frequencies are greater than *cis*-influences. (A great deal of evidence for this assumption is presented in Chapter 7, and theoretical rationalisations are given in Chapter 6.) Thus, if the mean stretching parameters of systems containing a ligand L' are lower than those of analogous systems containing a ligand L, the value of δ in complexes of the type *fac*-$M(CO)_3 L_2L'$ is expected to be negative, since the unique CO group is trans to L'.

It may sometimes seem possible to distinguish between the two assignments from the values generated for i; one assignment may lead to values for i that are unreasonable when compared with those for model compounds, or even to unreal roots. This situation can, however, only arise if δ is not small. In such cases the fault may lie not in a particular assignment, but in the assumption that all the interaction constants are equal. The molecule, whichever assignment is correct, exemplifies Case (iii) discussed below.

A valid procedure for distinguishing between the two possibilities depends on careful examination of the ^{13}CO satellite bands. Substitution by ^{13}CO may occur either at CO(1) or at CO(2), CO(3). In the former case, the symmetry of the molecule is not lowered from C_3, so that $v(A'')$ is unaffected. Both the A' bands are affected, to extents depending on how much the distortion of CO(1) contributes to their normal modes. Substitution at CO(3) will destroy the molecular plane of symmetry and affect all three modes. Thus the A'' band will have only one satellite band, while the A' bands will each have two, with intensity ration 2:1. Moreover, the Teller–Redlich product rule will be obeyed by the parent peaks and weaker satellites of the A' bands alone. To apply this criterion, it is, however, necessary to locate both the weaker satellite bands.

A third possible way of distinguishing between the A' and A'' modes is by depolarisation ratio measurements in the Raman spectrum. This is likely to be difficult, since for small δ the lower A' mode approximates closely to one of the E modes of Case (i), and will thus be almost fully depolarised despite its high formal symmetry.

The ambiguity that arises from the quadratic nature of Eq. (3.50) is serious only for moderate positive values of δ. The form of Eq. (3.50) is such that the two possible solutions are related by an interchange of the values of δ and i. For mononuclear carbonyls, i may be presumed positive, and approximate values of i may be inferred from model compounds. In some cases, however, a genuine ambiguity may persist. This can, in favourable circumstances, be resolved by inspection of the weaker ^{13}CO satellite bands, which should fit the modified secular equation

$$\begin{vmatrix} k + 2\delta - K & \sqrt{2}iR(^{13}CO) \\ \sqrt{2}iR(^{13}CO) & (k - \delta + i)R^2\,(^{13}CO) - K \end{vmatrix} = 0. \qquad (3.54)$$

Case (iii): in species of type fac-M(CO)$_3$ L$_2$ L$'$, the distinction between the two kinds of interaction constant cannot be ignored if L and L are very different. Symmetry coordinates may be chosen according to Eq. (3.50), and the secular equations may then be written in the form

$$A': \begin{vmatrix} k + 2\delta - K & \sqrt{2}i' \\ \sqrt{2}i' & k - \delta + i - K \end{vmatrix} = 0$$

$$A'': K = k - i - \delta. \qquad (3.55)$$

In the Cotton–Kraihanzel approximation, Eq. (3.55) reduces to Eq. (3.49), and Case (iii) reduces to Case (ii). This approximation may or may not be acceptable for any given molecule.

The complete range of possible sets of parameters may be found by the "independent variable" method of Section 3.4. The mean stretching parameter k takes the fixed value $[K(A')_1 + K(A')_2 + K(A'')]/3$. Rearrangement of Eq. (3.55) shows that the remaining parameters may conveniently be expressed in terms of δ:

$$i = k - \delta - K(A'')$$

$$([K(A')_1 - K(A')_2]/4)^2 = (\delta - [K(A')_1 + K(A')_2 - 2k]/4)^2 + (i'/\sqrt{2})^2.$$

$$(3.56)$$

The plot of $i'/\sqrt{2}$ against δ is a circle with a radius of $[K(A')_1 - K(A')_2]$, centred at the point $(\delta = [K(A')_1 + K(A')_2 - 2k]/4, i'/\sqrt{2} = 0)$.

Case (iv): The full treatment of species mer-M(CO)$_3$L$_3$, M(CO)$_3$L (square planar), or M(CO)$_3$L$_2$ (equatorially substituted trigonal bipyramid),

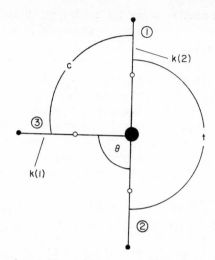

FIG. 3.7. Energy factored force field for *mer*-M(CO)$_3$L$_3$ and related species.

all of symmetry C_{2v}, is closely related to that of the preceding paragraph. In the notation of Fig. 3.7 the symmetry coordinates may be chosen as

$$S(A_1)_1 = [\Delta(1) + \Delta(2)]/\sqrt{2}$$
$$S(A_1)_2 = \Delta(3)$$
$$S(B_2) = [\Delta(1) - \Delta(2)]/\sqrt{2} \qquad (3.57)$$

and the secular equations are

$$A_1: \begin{vmatrix} k(2) + t - K & \sqrt{2}c \\ \sqrt{2}c & k(1) - K \end{vmatrix} = 0$$

$$B_2: K = k(2) - t \qquad (3.58)$$

The force field is underdetermined, with one degree of freedom. If $k(1)$ is selected as the independent variable, the other parameters are given by the equations

$$k(2) = [K(A_1)_1 + K(A_1)_2 + K(B_2) - k(1)]/2$$
$$t = [K(A_1)_1 + K(A_1)_2 - K(B_2) - k(1)]/2$$
$$[K(A_1)_1 - K(A_1)_2]^2 = [2k(1) - K(A_1)_1 - K(A_1)_2]^2 + (2\sqrt{2}c)^2$$

$$(3.59)$$

The plot of $\sqrt{2}\,c$ against $k(1)$ is a circle with a radius of $[K(A')_1 - K(A')_2]/2$, centred at the point $k(1) = [K(A')_1 + K(A')_2]/2$, $\sqrt{2}c = 0$. The Cotton–Kraihanzel approximation takes $c = i$, $t = 2i$, leading to the particular solution

$$9k(1)^2 - k(1)\,[10K(A_1)_1 + 10K(A_1)_2 - 2K(B_2)]$$
$$+ [K(A_1)_1 + K(A_1)_2 - K(B_2)]^2 + 8K(A_1)_1\,K(A_1)_2 = 0$$

$$i = K(A_1)_1 + K(A_1)_2 - K(B_2) - k(2)]/4$$

$$k_2 = K(B_2) + 2i \qquad\qquad (3.60)$$

The smaller of the two possible values for $k(1)$ must be chosen, since $k(1)$ is smaller than $k(2) + 2i$. The particular solution given by Eq. (3.60) is not in any way privileged, and may not be real.

The assignment of the three infrared bands requires care. The highest frequency corresponds to $(A_1)_1$. The order of $(A_1)_2$ and B_2 depends on the sizes of the interaction constants and of the difference between $k(1)$ and $k(2)$, and is variable. Were the local oscillating dipoles of the two kinds of carbonyl group equal, then the B_2 mode would be expected to have the higher integrated intensity by a factor of around two. However, the oscillating dipole of CO(3) is likely to be higher than those of CO(1, 2), so that this criterion may fail. More reliable is the combined use of infrared and Raman spectroscopy. Although all three bands are both Raman and infrared allowed, the B_2 mode is expected to be weak in the Raman spectrum since the changes in polarisability of the two CO groups involved in this mode will very nearly cancel. It will also, of course, be fully depolarised.

The forms of the normal modes (Fig. 3.8) and the relative intensities may

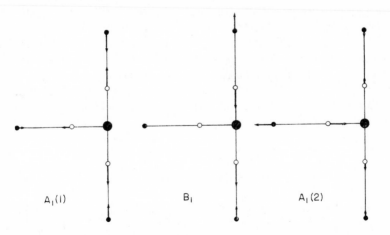

FIG. 3.8. Normal CO stretching modes of *mer*-M(CO)$_3$L$_3$ and related species.

most readily be found using the $\cos 2\phi$ technique of Section 3.5. In this case, Eqs (3.38) and (3.39) become

$$k(1) = [K(A_1)_1 (1 - \cos 2\phi) + K(A_1)_2 (1 + \cos 2\phi)]/2$$

$$k(2) = [K(A_1)_1 (1 + \cos 2\phi) + K(A_1) (1 - \cos 2\phi)]/4 + K(B_2)/2$$

$$c = [K(A_1)_1 - K(A_1)_2] \sin 2\phi/2\sqrt{2} \tag{3.61}$$

and

$$S(A_1)_1 = [\Delta(1) + \Delta(2)]/\sqrt{2} = \mu^{-\frac{1}{2}}[Q(A_1)_1 \cos \phi + Q(A_1)_2 \sin \phi]$$

$$S(A_1)_2 = \Delta(3) = \mu^{-\frac{1}{2}}[- Q(A_1)_1 \sin \phi + Q(A_1)_2 \cos \phi]$$

$$S(B_2) = [\Delta(1) - \Delta(2)]/\sqrt{2} \tag{3.62}$$

The relative intensities of the three modes may then be inferred using Eqs (3.31, 3.36):

$$I(A_1)_1 : I(A_1)_2 : I(B_2) :: [2\mu(1) \cos \theta(c) \cos \phi + \mu(3) \sin \phi]^2 :$$

$$[- 2\mu(1) \cos \theta(c) \sin \phi + \mu(3) \cos \phi]^2 :$$

$$[2\mu(1) \sin \theta(c)]^2. \tag{3.63}$$

If CO(1) and CO(2) are exactly *trans*, so that $\theta(t)$ is 180°, and the interaction coefficients $\sin \phi$ can be ignored, then the higher A_1 band would be vanishingly weak in the infrared spectrum. Both geometric and coupling effects are important in determining the actual intensity of this band. Since mutually *cis* carbonyl groups appear to exert a strong mutual steric repulsion, $\theta (c)$ is likely to exceed 90°. At the same time, $\sin \phi$ is positive, so that the geometric and coupling contributions to the oscillating dipole are opposed (Fig. 3.8), and the higher A_1 peak may be extremely weak even though neither of the two contributions are negligible.

If the force field is known, it is possible to use infrared intensity data to estimate $\mu(1)$, $\mu(3)$ and $\theta(c)$; and even if the force field is not known, $\mu(1)$ and $\mu(3)$ may be estimated by neglecting small terms:

$$I(A_1)_2 \doteq [N\pi/3c^2] [\mu(3)]^2$$

$$I(B_2) \doteq 2[N\pi/3c^2] [\mu(1)]^2. \tag{3.64}$$

It is also possible to estimate $\cos \theta(c)$ in terms of $\tan \phi$, and hence to find $\theta(c)$ as a function of the force field. This calculation can be carried out using the approximations of Eq. (3.64), or, slightly more precisely, using Eq. (3.63).

M(CO)₄

Case (i): the treatment of tetrahedral species $M(CO)_4$ is straightforward. The point-group is T_d and the representations spanned are A_1 and T_2. The forms of the symmetry modes are Fig. 3.9

$$S(A_1) = [\Delta(1) + \Delta(2) + \Delta(3) + \Delta(4)]/2$$

$$S(T_2)_x = [\Delta(1) + \Delta(2) - \Delta(3) - \Delta(4)]/2$$

$$S(T_2)_y = [\Delta(1) - \Delta(2) + \Delta(3) - \Delta(4)]/2 \qquad (3.65)$$

$$S(T_2)_z = [\Delta(1) - \Delta(2) - \Delta(3) + \Delta(4)]/2$$

and the secular equations are

$$A_1 : K = k + 3i$$

$$T_2 : K = k - i. \qquad (3.66)$$

The A_1 mode is Raman-active and fully polarised, since the anisotropy is zero; it is infrared-inactive. The T_2 mode is Raman-active (depolarised) and infrared-active. If the Raman spectrum is unobtainable, the A_1 frequency may be estimated from the spectrum of $M(^{12}CO)_3\ ^{13}CO$, which occurs naturally at $4\frac{1}{2}\%$ abundance. This species belongs to the point-group C_{3v}, and the secular equations are:

$$A_1: \begin{vmatrix} k + 2i - K & R\sqrt{3}i \\ R\sqrt{3}i & R^2 k - K \end{vmatrix} = 0$$

$$E: K = k - i. \qquad (3.67)$$

The E modes are linear combinations of the T_2 modes of the parent species, while the A_1 modes correlate with the parent A_1 mode and the third linear combination of T_2 modes. If both the A_1 modes of the labelled species are observed, then it is possible to find the missing mode of the parent species by using the Teller–Redlich product rule:

$$v(A_1)_1\ [M(CO)_3\ (CO)^*]\ v(A_1)_2\ [M(CO)_3\ (CO)^*]/v(A_1)\ [M(CO)_4]$$

$$\times\ v(T_2)\ [M(CO)_4] = R. \quad (3.68)$$

Case (ii): if one CO group is slightly different from the others, so that the differences in interaction parameter can be ignored, then the stretching parameters of the unique CO and the three others may be assigned values

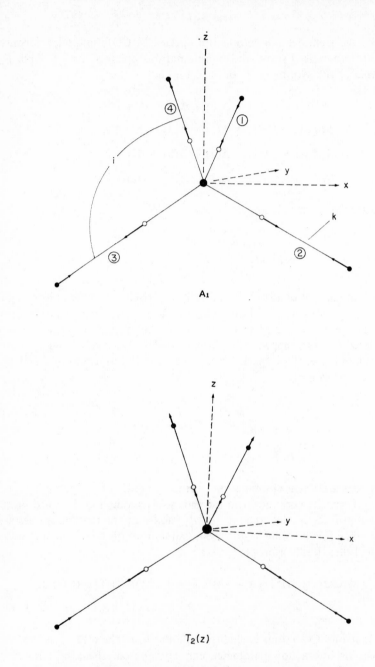

FIG. 3.9. Force field and symmetry coordinates of tetrahedral $M(CO)_4$.

$k + 3\delta$ and $k - \delta$ respectively. Taking the unique CO group as CO(1), the symmetry modes in C_{3v} become

$$S(A_1)_1 = [\Delta(2) + \Delta(3) + \Delta(4)]/\sqrt{3}$$

$$S(A_1)_2 = \Delta(1)$$

$$S(E) \;\;= [\Delta(2) - \Delta(3)]/\sqrt{2}$$

$$S'(E) \;\;= [2\Delta(4) - \Delta(2) - \Delta(3)]/\sqrt{6} \tag{3.69}$$

giving secular equations

$$A_1: \begin{vmatrix} k + 2i - \delta - K & \sqrt{3}i \\ \sqrt{3}i & k + 3\delta - K \end{vmatrix} = 0$$

$$E: K = k - i - \delta. \tag{3.70}$$

The treatment proceeds much as for Case (ii) of $M(CO)_3$. Ignoring terms in δ^2 gives solutions

$$K(A_1)_1 = k + 3i$$

$$K(A_1)_2 = k - i + 2\delta$$

$$K(E) \;\;= k - i - \delta. \tag{3.71}$$

Thus, in the limit of δ being very much smaller than i, the fully symmetric mode is unaffected by the asymmetry under discussion. The asymmetric mode is split into two peaks with intensity ratio 2:1, and it is possible to determine the sign of δ from the frequency ordering of these two peaks.

Since the expanded form of Eq. (3.70) is symmetrical in i and δ, the solutions for $\delta \gg i$ become

$$K(A_1)_1 = k + 3\delta$$

$$K(A_1)_2 = k + 2i - \delta$$

$$K(E) \;\;= k - i - \delta. \tag{3.72}$$

The dependence of the values of K on δ is shown in Fig. 3.10. There is no ambiguity in the assignments (though the $(A_1)_1$ mode will be weak in the infrared spectrum) or in the sign of δ. There is, however, an ambiguity of choice between solutions in which δ and i are interchanged, which can (if both are positive) only be resolved by reference to model systems or by analysis of the spectrum of isotopically labelled species.

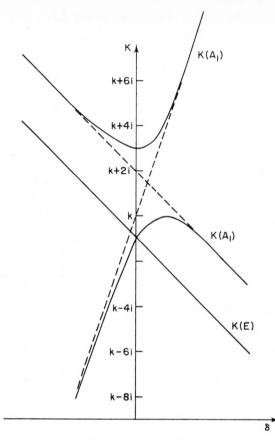

FIG. 3.10. Behaviour of the vibrations of $M(CO)_4$, Case (ii), as a function of δ.

Case (*iii*) arises if the unique CO group is so markedly different from the others that the interaction parameters cannot be presumed to be identical. For example, a tight ion pair such as $\left(M(CO)_4{}^-\right)(M')^+$ may exemplify Case (ii), while a compound of type $(OC)_3\, M(CO) \rightarrow M'$, in which the unique carbonyl group is bifunctional, would belong to Case (iii). The secular equations are

$$A: \begin{vmatrix} k + 2i - \delta - K & \sqrt{3}i' \\ \sqrt{3}i' & k + 3\delta - K \end{vmatrix} = 0 \qquad (3.73)$$

$$E: K = k - i - \delta.$$

The possible solutions may be displayed by choosing δ as an independent variable. The mean stretching parameter k takes the fixed value

$[K(A_1)_1 + K(A_1)_2 + 2K(E)]/4$. Rearrangement of Eq. (3.73) gives

$$i = k - \delta - K(E)$$

$$[K(A_1)_1 - K(A_1)_2]^2 = \{6\delta - [K(A_1)_1 + K(A_1)_2 - 2k]\}^2 + (2\sqrt{3}i')^2. \quad (3.74)$$

The second part of Eq. (3.74) shows that a plot of $i'/\sqrt{3}$ against δ is a circle of radius $[K(A_1)_1 - K(A_1)_2]/6$, with its centre at the point $\{\delta = [K(A_1)_1 + K(A_1)_2 - 2k]/6, i'/\sqrt{3} = 0\}$.

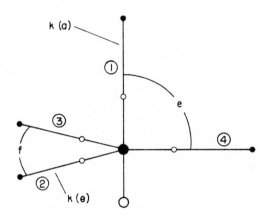

FIG. 3.11. Force field of trigonal bipyramidal M(CO)₄L.

An important class of compounds to which the formal theory of this passage applies is that of axially substituted trigonal bipyramidal species M(CO)₄L Fig. 3.11. For these, Eqs (3.73, 3.74) may be rewritten as

$$A: \begin{vmatrix} k(e) + 2f - K & \sqrt{3}e \\ \sqrt{3}e & k(a) - K \end{vmatrix} = 0$$

$$E: K = k(e) - f \quad (3.75)$$

and

$$k(e) = [K(A_1)_1 + K(A_1)_2 + 2K(E) - k(a)]/3$$

$$f = k(e) - K(E)$$

$$[K(A_1)_1 - K(A_1)_2]^2 = \{2k(a) - [K(A_1)_1 + K(A_1)_2]\}^2 + (2\sqrt{3}e)^2. \quad (3.76)$$

The plot of $\sqrt{3}e$ against $k(a)$ is a circle, of radius $[K(A_1)_1 - K(A_1)_2]/2$ with its centre at the point $\{k(a) = [K(A_1)_1 + K(A_1)_2]/2, \sqrt{3}e = 0\}$.

The forms of the normal modes, and the relative intensities, are more readily found by the $\cos 2\phi$ method of Eqs (3.38, 3.39). It is possible to choose ϕ so that

$$k(e) = [K(A_1)_1 (1 + \cos 2\phi) + K(A_1)_2 (1 - \cos 2\phi)]/6 + K(E)/3$$
$$k(a) = [K(A_1)_1 (1 - \cos 2\phi) + K(A_1)_2 (1 + \cos 2\phi)]/2$$
$$e = [K(A_1)_1 - K(A_1)_2] \sin 2\phi/2\sqrt{3}. \tag{3.77}$$

Then the normal modes are as shown in Fig. 3.11a.

$$S(A_1)_e = [\Delta(1) + \Delta(2) + \Delta(3)]/\sqrt{3}$$
$$= \mu^{-\frac{1}{2}} [\cos\phi \cdot Q(A_1)_1 + \sin\phi \cdot Q(A_1)_2]$$
$$S(A_1)_a = \Delta(4)$$
$$= \mu^{-\frac{1}{2}} [-\sin\phi \cdot Q(A_1)_1 + \cos\phi \cdot Q(A_1)_2] \tag{3.78}$$

and the calculated relative intensities are given by

$$I(A_1)_1 : I(A_1)_2 : I(E) :: [3\cos\phi\cos\theta(e)\mu(a) + \sin\phi\,\mu(e)]^2 :$$
$$[- 3\sin\phi\cos\theta(e)\mu(a) + \cos\phi\,\mu(e)]^2 : [3\sin\theta(e)\mu(a)]^2. \tag{3.79}$$

As in Case (iv) of $M(CO)_3$, the geometric and coupling contributions to the intensities of the highest modes are opposed. The value of $\mu(e)$ may be

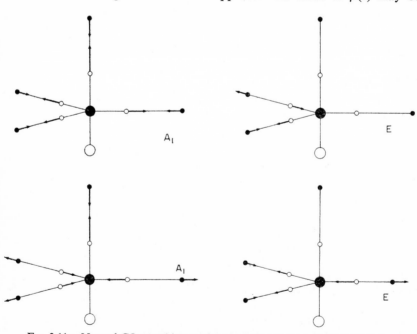

Fig. 3.11a. Normal CO stretching modes of trigonal bipyramidal $M(PCO)_4L$.

estimated from the intensity of the E mode, assuming $\sin \theta(e)$ equal to one, but unless the coupling coefficient $\sin \phi$ is taken to be small it is not possible to estimate $\mu(a)$ without first estimating ϕ.

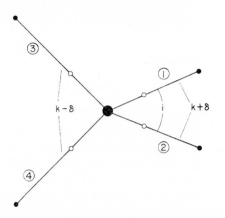

FIG. 3.12. Force field of perturbed tetrahedral $M(CO)_4$, Case (iv).

Case (iv): if two CO groups of Case (i) are perturbed equally (for example, by the incorporation of an ionic tetrahedral carbonyl into an ion pair of symmetry C_{2v}) the symmetry coordinates may be chosen (Fig. 3.12) as

$$S(A_1)_1 = [\Delta(1) + \Delta(2)]/\sqrt{2}$$
$$S(A_1)_2 = [\Delta(3) + \Delta(4)/\sqrt{2}$$
$$S(B_1) = [\Delta(1) - \Delta(2)]/\sqrt{2}$$
$$S(B_2) = [\Delta(3) - \Delta(4)]/\sqrt{2}. \tag{3.80}$$

The secular equations are

$$A_1: \begin{vmatrix} k + i + \delta - K & 2i \\ 2i & k + i - \delta - K \end{vmatrix} = 0 \tag{3.81}$$

$$B_1: K = k - i + \delta$$

$$B_2: K = k - i - \delta.$$

These equations imply a frequency ordering $(A_1)_1 > B_1 > (A_1)_2 > B_2$, the relative labels of B_1 and B_2 being chosen so that δ is positive. In the limit of

δ being very much smaller than i, the roots to the secular equations become

$$K(A_1)_1 = k + 3i$$
$$K(A_1)_2 = k - i$$
$$K(B_1) \ = k - i + \delta$$
$$K(B_2) \ = k - i - \delta.$$

(3.82)

In the limit of δ being very much greater than i, the roots for the fully symmetric vibration become

$$K(A_1)_1 = k + i + \delta$$
$$K(A_1)_2 = k + i - \delta$$

(3.83)

while the expressions for $K(B_1)$ and $K(B_2)$ are unaltered. The behaviour of the effective force constants as a function of δ is shown in Fig. 3.13. It is noteworthy that the frequency ordering is the same whatever the (positive) value of δ. It is also of interest that, provided δ is large enough for the splitting of the T_2 mode of Case (i) to be detectable, this spliting should give rise to three resolvable bands; thus there is no risk of confusion between Cases (ii) and (iv). All four bands of Case (iv) are active in both infrared and Raman spectra; but if the environmental perturbation is small, the higher A_1 band may be weak in the infrared, and the lower A_1 band of the Raman spectrum may be almost fully depolarised.

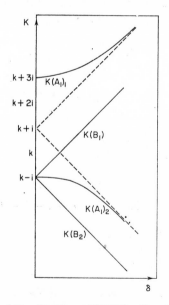

FIG. 3.13. Behaviour of the vibrations of M(CO)$_4$, Case (iii), as a function of δ.

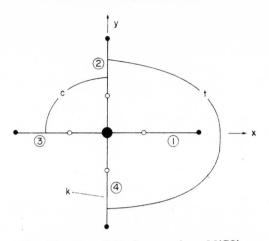

FIG. 3.14. Force field of square planar $M(CO)_4$.

Case (v): the treatment of square planar $M(CO)_4$ (Fig. 3.14) is straightforward in principle, but this is true in practice only if Raman data are obtainable. The point-group is D_{4h}, and the forms of the symmetry coordinates (Fig. 3.15) are

$$S(A_{1g}) = [\Delta(1) + \Delta(2) + \Delta(3) + \Delta(4)]/2$$

$$S(B_{2g}) = [\Delta(1) - \Delta(2) + \Delta(3) - \Delta(4)]/2$$

$$S(E_u)_x = [\Delta(1) - \Delta(3)]/\sqrt{2}$$

$$S(E_u)_y = [\Delta(2) - \Delta(4)]/\sqrt{2}. \tag{3.84}$$

(The second mode listed is sometimes assigned to the representation B_{1g}; the distinction arises from a choice in the naming of the symmetry operations of the point-group and is without significance. Any set of orthogonal normalised combinations of $S(E_u)_x$ and $S(E_u)_y$ may be used to span the representation E.)

The secular equations are

$$K(A_{1g}) = k + t + 2c$$

$$K(B_{2g}) = k + t - 2c$$

$$K(E_u) = k - t. \tag{3.85}$$

There are three unknowns and three frequencies so that the problem is determinate. The frequency ordering is $v(A_{1g}) > v(B_{2g})$ and, since t is always found to be greater than c, then $v(B_{2g}) > v(E_u)$. The two higher

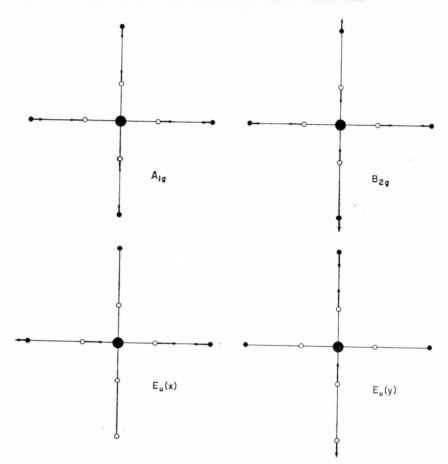

FIG. 3.15. Symmetry coordinates of square planar $M(CO)_4$.

frequencies are Raman-active but infrared-inactive, while the lowest frequency is active in the infrared spectrum only.

If Raman data are not available, the force field can in principle be found by isotopic labelling. If we choose the labelled CO group of the singly labelled species to be CO(1), then symmetry coordinates may be chosen in C_{2v} as

$$S(A_1)_1 = [\Delta(2) + \Delta(4)]/\sqrt{2}$$
$$S(A_1)_2 = \Delta(3)$$
$$S(A_1)_3 = \Delta(1)$$
$$S(B_2) \ = [\Delta(2) - \Delta(4)]/\sqrt{2}. \tag{3.86}$$

The secular equations then are

$$
A_1: \begin{vmatrix} k + t - K & \sqrt{2}c & R\sqrt{2}c \\ \sqrt{2}c & k - K & Rt \\ R\sqrt{2}c & Rt & R^2k - K \end{vmatrix} = 0 \qquad (3.87)
$$

$$B_2: \quad K = k - t.$$

The B_2 mode is identical with an E mode of the unlabelled species and gives no new information. The three A_1 modes are all in principle infrared active. The lowest mode will be readily observable as a low-frequency satellite to the parent E_u band even at natural ($4\frac{1}{2}\%$) abundance of the labelled species. There will be two higher frequency bands due to the isotopically labelled species, and the positions and relative intensities of these will depend on the relative magnitudes of t and c. The presence of two high energy isotope bands would therefore confirm a square planar rather than a tetrahedral arrangement, but the observation of one band only is, strictly speaking, inconclusive. The A_1 modes of the labelled square planar species correlate with both the A_{1g} and B_{2g} modes of the parent, as well as with one of the E_x modes. It is thus not possible to gain any information from using the Teller–Redlich product rule, since there are two unknowns occurring in the one equation. It is, however, possible in principle to find t, c and k. Of these, k and t may be fixed using the properties of the sum of the roots of a determinant:

$$
K(A_1)_1 + K(A_1)_2 + K(A_1)_3 = k(2 + R^2) + t \qquad (3.88)
$$

together with the expression for $K(B_2)$ in Eq. (3.87). Substituting in the determinant of Eq. (3.87) and expanding then gives a quadratic equation for c; this has two roots, one of which is negative and may be rejected.

Case (vi): if a species $M(CO)_4$ is intermediate between tetrahedral and square planar forms, then it belongs to the point-group D_{2d}. The theoretical treatment follows that of Case (v) very closely, differing only in the labelling of the symmetry modes and in the selection rules. The A_{1g}, B_{2g} and E_u modes of D_{4h} are re-labelled A_1, B_2 and E respectively. All three modes are formally Raman-active, although if the species is close to planarity the Raman intensity of the E mode will be small. The A_1 band remains formally infrared-inactive but the B_2 mode is active, and in the notation of Fig. 3.16 the intensity ratio of the two allowed bands is given by

$$
I(B_2)/I(E) = \tan^2 \omega. \qquad (3.89)
$$

It has been suggested that octahedral species of the type *trans*-$M(CO)_4L_2$ show distortion from the regular arrangement of Case (v) to that of Case (vi).

Such distortion has two important consequences. Firstly, the B_{2g} mode acquires some infrared-allowed character. Secondly, the distinction between t and c must be reduced, since extreme distortion of a square planar complex converts it to a tetrahedron, in which the distinction vanishes. The separation between $k(B_2)$ and $k(E)$ is $2t - 2c$. Distortion away from the square planar configuration reduces this quantity, which becomes zero in the tetrahedral limit as the B_2 and E species of D_{2d} become different sub-species of T_2 in T_d.

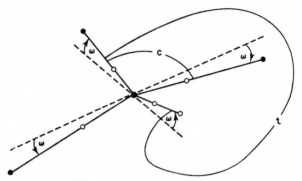

FIG. 3.16. Force field of M(CO)$_4$ intermediate between square planar and tetrahedral geometry.

Case (vii): it is possible for an array of four carbonyl ligands to belong to the point-group C_{4v}. This situation arises in species of type *trans*-M(CO)$_4$LL′, but the deviation from planarity does not seem likely to be significant. More significant is the deviation from co-planarity expected for the carbonyl groups of $(\pi$-$C_5H_5)V(CO_4)$.

The forms of symmetry coordinates and the secular equations are identical with D_{4h} [case (iv)] except that the labels g and u are of course dropped. The A_1 mode becomes infrared allowed, with

$$I(A_1)/I(E) = 2\tan^2 \omega. \tag{3.90}$$

(Here ω is the angle by which the four CO groups deviate from a plane perpendicular to the four fold axis.) All three modes are in principle Raman-allowed, but, unless ω is appreciable, the Raman intensity of the E mode will be low.

Case (viii): typified by species *cis*-M(CO)$_4$L$_2$ or L(*axial*) Fe(CO)$_4$; see Appendix IV.

M(CO)$_5$

There are two common arrangements of five carbonyl groups around a metal. The first of these, the trigonal biprism (Cases (i) and (ii)), is found in simple carbonyls. The second, a C_{4v} arrangement (Cases (iii) and (iv)), occurs in species of the type $M(CO)_5L$. Both arrangements belong to carbonyls of the second type.

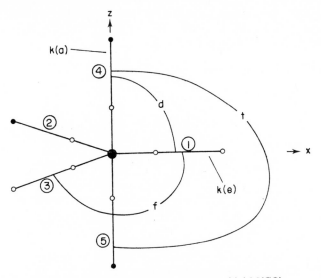

FIG. 3.17. Force field of trigonal bipyramidal $M(CO)_5$.

Case (i): in iron pentacarbonyl and related species, the arrangement of the carbonyl groups around the metal atom is trigonal bipyramidal (point-group D_{3h}). The secular equations are, in the notation of Fig. 3.17,

$$A_1' : \begin{vmatrix} k(e) + 2f - K & \sqrt{6}d \\ \sqrt{6}d & k(a) + t - K \end{vmatrix} = 0$$

$$A_2'' : K = k(a) - t$$

$$E' : K = k(e) - f. \tag{3.91}$$

The force field is underdetermined, with one degree of freedom. Choosing $k(e)$ as the independent variable gives

$$K(A_1')_1 + K(A_1')_2 + K(A_2'') + 2K(E') = 2k(a) + 3k(e)$$

$$[K(A_1')_1 - K(A_1')_2]^2 = \{6k(e) - [K(A_1')_1 + K(A_1')_2 + 4k(E')]\}^2 + 24d^2. \tag{3.92}$$

Thus, a plot of $\sqrt{2/3}\, d$ against $k(e)$ is a circle of radius $(K(A_1')_1 - K(A_1')_2)/6$, centred at the point $\{k(e) = [K(A_1')_1 + K(A_1')_2 + 4E']/6, \sqrt{2/3}\, d = 0.\}$

The forms of the normal modes are most readily found by the $\cos 2\phi$ method of Eqs (3.38, 3.39). It is possible to choose ϕ so that

$$k(e) = [K(A_1')_1 (1 + \cos 2\phi) + K(A_1')_2 (1 - \cos 2\phi)]/6 + K(E')/3$$
$$k(a) = [K(A_1')_1 (1 - \cos 2\phi) + K(A_1')_2 (1 + \cos 2\phi)]/4 + K(A_2'')/2$$
$$d = [K(A_1')_1 - K(A_1')_2]\sin \phi/2\sqrt{6}. \tag{3.93}$$

Then the normal modes are as shown in Fig. 3.18:

$$\begin{aligned}
S(A_1')_e &= [\Delta(1) + \Delta(2) + \Delta(3)]/\sqrt{3} \\
&= \mu^{-\frac{1}{2}} [Q(A_1')_1 \cos \phi + Q(A_1')_2 \sin \phi] \\
S(A_1')_a &= [\Delta(4) + \Delta(5)]/\sqrt{2} \\
&= \mu^{-\frac{1}{2}} [- Q(A_1')_1 \sin \phi + Q(A_1')_2 \cos \phi] \\
S(A_2'') &= [\Delta(4) - \Delta(5)]/\sqrt{2} \\
S(E')_x &= [2\Delta(2) - \Delta(3) - \Delta(4)]/\sqrt{6} \\
S(E')_y &= [\Delta(2) - \Delta(3)]/\sqrt{2}.
\end{aligned} \tag{3.94}$$

The A_1' modes are infrared inactive. The A_2'' and E' modes are infrared active, with relative intensities

$$I(A_2'')/I(E') = 2\mu(a)^2/3\mu(e)^2. \tag{3.95}$$

The A_1' modes are Raman-active, as is the E' mode, while the A_2'' mode is Raman-inactive. An unambiguous assignment of the infrared spectrum is not possible without Raman or other additional data, but (as the discussion of Case (ii) shows) solvent broadening data may suffice.

At natural abundance, isotopic labelling will occur in axial and equatorial positions, in $2 \cdot 2\%$ and $3 \cdot 3\%$ of molecules respectively. The axially substituted species belongs to the point-group C_{3v}, and the secular equation for the A modes is

$$\begin{vmatrix} k(e) + 2f - K & \sqrt{3}d & R\sqrt{3}d \\ \sqrt{3}d & k(a) - K & Rt \\ R\sqrt{3}d & Rt & R^2 k(a) - K \end{vmatrix} = 0. \tag{3.96}$$

The E modes are identical with the E' modes of the unlabelled parent species.

The equatorially substituted species belongs to the point-group C_{2v}. The A_2'' and E_y' modes of the parent species are re-labelled B_1 and B_2, but

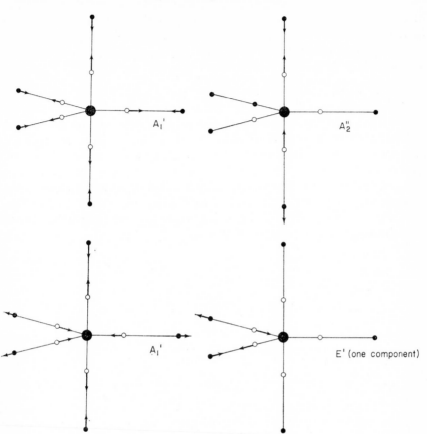

FIG. 3.18. Normal CO stretching modes of trigonal bipyramidal $M(CO)_5$ (only one component of E' shown).

are otherwise unchanged. The remaining modes belong to the representation A_1, and the secular equation is

$$\begin{vmatrix} k(e) + f - K & 2d & R\sqrt{2}f \\ 2d & k(a) + t - K & R\sqrt{2}d \\ R\sqrt{2}f & R\sqrt{2}d & R^2 k(e) - K \end{vmatrix} = 0. \qquad (3.97)$$

Thus, there will be a total of six isotope bands in the infrared spectrum of trigonal bipyramidal $M(CO)_4 (CO)^*$. Two of these will be readily identifiable as satellites of the two infrared bands, while the remainder will be at higher frequency, derived mainly from inactive modes, and correspondingly weaker. It is not possible to use the Teller–Redlich product rule to find the infrared inactive frequencies of the parent species, since these both belong to the same, fully symmetric, representation. It will, however, be possible to infer the frequencies from the force field, the independent variable of which can be chosen to give the best fit for the available data.

Case (ii): slight deviations from Case (i) may be expected for ionic species involved in ion pair formation, or for any species interacting with solvent. A deviation affecting the axial CO groups only, may be represented by a shift of the stretching parameters of CO(4) and CO(5) to values $k(a) + \delta$ and $k(a) - \delta$ respectively. Applying perturbation theory then shows that if δ is far smaller than t, the effect on $K(A_1')_1$, $K(A_1')_2$ and $K(A_2'')$ is proportional to $(\delta/t)^2$, and is zero to first order. A deviation affecting the equatorial groups only, can be discussed using the theory of $M(CO)_3$, Case (ii); the effect to first order is to cause symmetric splitting of the E mode while leaving the other modes unaffected.

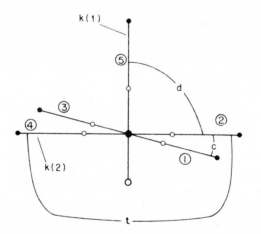

FIG. 3.19. Force field of square pyramidal $M(CO)_5$.

Case (iii): in a very large number of species of type $M(CO)_5L$, as also in photolytic fragments of type $M(CO)_5$ [43, 44, 45], the arrangement of the carbonyl groups is square pyramidal (C_{4v}), Fig. 3.19. The secular equations are

$$A_1 : \begin{vmatrix} k(2) + t + 2c - K & 2d \\ 2d & k(1) - K \end{vmatrix} = 0$$

$$B_2 : K = k(2) + t - 2c$$

$$E : K = k(2) - t. \tag{3.98}$$

The A_1 and E modes are infrared active, and also formally Raman active, though the E modes resemble in form the T_{1u} modes of $M(CO)_6$ (see below), and are expected to be weak. The B_2 mode is formally active in the Raman spectrum only, although small distortions (Case (iv) below) can impart infrared activity.

The force field is underdetermined, with one degree of freedom. Choosing $k(1)$ as the independent variable gives

$$[K(A_1)_1 - K(A_1)_2]^2 = [2k(1) - K(A_1)_1 - K(A_1)_2]^2 + (4d)^2$$

$$k(2) = [K(A_1)_1 + K(A_1)_2 + K(B_2) + 2K(E) - k(1)]/4$$

$$t = k(2) - K(E)$$

$$c = [k(2) + t - K(B_2)]/2. \tag{3.99}$$

Thus, the plot of $2d$ against $k(1)$ is a circle of radius $[K(A_1)_1 - K(A_1)_2]/2$, centered at the point $\{k(1) = [K(A_1)_1 + K(A_1)_2]/2, d = 0\}$, and $k(1)$ is constrained to lie between $K(A_1)_1$ and $K(A_1)_2$. In addition, it is expected on general grounds [34] that (except perhaps where CO(5) is *trans* to PF$_3$, or a similar ligand) $k(2) > k(1)$; this may in some cases further restrict the range of possible values of $k(1)$.

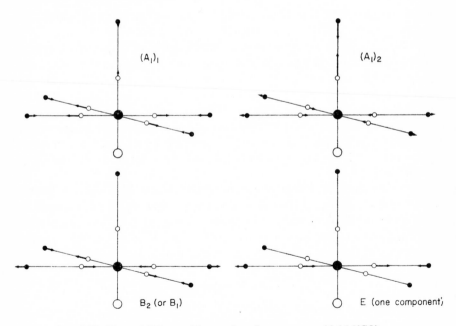

$(A_1)_1$ $(A_1)_2$

B_2 (or B_1) E (one component)

FIG. 3.20. Normal CO stretching modes of square pyramidal $M(CO)_5$.

The force field may also be described using the $\cos 2\phi$ method of Eqs (3.38, 3.39):

$$k(2) = [K(A_1)_1 (1 + \cos 2\phi) + K(A_1)_2 (1 - \cos 2\phi)]/8 + K(B_2)/4 + K(E)/2$$
$$k(1) = [K(A_1)_1 (1 - \cos 2\phi) + K(A_1)_2 (1 + \cos 2\phi)]/2$$
$$2d = [K(A_1)_1 - K(A_1)_2] \sin 2\phi/2. \tag{3.100}$$

The forms of the normal modes are then given Fig. 3.20 by

$$S(A_1)_e = [\Delta(1) + \Delta(2) + \Delta(3) + \Delta(4)]/2$$
$$= \mu^{-\frac{1}{2}} [Q(A_1)_1 \cos \phi + Q(A_1)_2 \sin \phi]$$
$$S(A_1)_a = \Delta(5)$$
$$= \mu^{-\frac{1}{2}} [- Q(A_1)_1 \sin \phi + Q(A_1)_2 \cos \phi]$$
$$S(B_2) = [\Delta(1) - \Delta(2) + \Delta(3) - \Delta(4)]/2 = \mu^{-\frac{1}{2}} Q(B_2)$$
$$S(E_x) = [\Delta(1) - \Delta(3)]/2 = \mu^{-\frac{1}{2}} Q(E_x)$$
$$S(E_y) = [\Delta(2) - \Delta(4)]/2 = \mu^{-\frac{1}{2}} Q(E_y). \tag{3.101}$$

The relative intensities of the infrared absorption bands are given by

$$I(A_1)_1 : I(A_1)_2 : I(E) :: [2 \cos \phi \cos \theta(d) \mu(2) + \sin \phi \, \mu(1)]^2 :$$
$$[- 2 \sin \phi \cos \theta(d) \mu(2) + \cos \phi \, \mu(1)]^2 : 4 \sin^2 \phi(d) \mu(1)^2. \tag{3.102}$$

There is no difficulty in most cases in assigning the infrared bands. The highest band is $(A_1)_1$ and is relatively weak, on occasion too weak to be found by superficial examination of a normal spectrum. The relative order of the $(A_1)_2$ and E bands depends on the particular values of the parameters, and on general grounds [40] $\mu(1)$ is expected to be greater than $\mu(2)$. Despite this it appears always to be the case that the strongest band in the spectrum belongs to the E mode, which also shows the greatest solvent broadening. The intensity of the highest A_1 mode arises from a balance of geometric and coupling effects. These are generally opposed (since $\theta(d)$ is commonly greater than 90°), and in some cases [21, 44] the resultant oscillating dipole is very near to zero.

If the B_2 mode cannot be observed in the infrared spectrum and Raman data are lacking, only three frequencies are available to fix five parameters. The force field problem is then underdetermined with two degrees of freedom, and it is possible to impose both constraints of the Cotton–Kraihanzel method (Eqs 3.24, 3.25). The secular equations of Eq. (3.98) then become

$$\begin{vmatrix} A_1 : k(2) + 4i - K & 2i \\ 2i & k(1) - K \end{vmatrix} = 0$$
$$E : K = k(2) - 2i \tag{3.103}$$

which may be rearranged to give

$$k(1) = K(A_1)_1 + K(A_1)_2 - K(E) - 6i$$

$$k(2) = K(E) + 2i$$

$$40i^2 - 6i[K(A_1)_1 + K(A_1)_2 - 2K(E)] +$$

$$[K(E) - K(A_1)_1][K(E) - K(A_1)_2] = 0. \tag{3.104}$$

The quadratic in i has two roots; a choice is made on the grounds that i must be positive and $k(2) + 4i$ must be greater than $k(1)$. Sometimes the CK force field found from Eq. (3.104) is used to calculate a position for an observable B band;

$$K(B_2)_{CK} = k(2). \tag{3.105}$$

Tolerable agreement is generally found, but this does not vindicate the CK force field, since there is no unique solution. In particular, it is not legitimate to claim any high degree of accuracy for the forms of normal modes calculated from CK parameters, or to make extended inferences, regarding relative infrared absorption intensities, from such calculations.

The indeterminacy of the force field of Eq. (3.98) can be removed using data for isotopically labelled species. At natural abundance, $1 \cdot 1\%$ of molecules will be axially substituted with ^{13}CO, while $4\frac{1}{2}\%$ will be substituted in one equatorial position. The axially substituted species retains the full symmetry of the parent. The secular equation for the A modes becomes

$$\begin{vmatrix} k(2) + t + 2c - K & 2Rd \\ 2Rd & R^2k(1) - K \end{vmatrix} = 0 \tag{3.106}$$

while the form of the B_2 and E modes is unaltered. Eq. (3.106) shows the existence of a weak band slightly to low frequency of the normal $(A_1)_1$ band of natural $M(CO)_5 L$, in addition to the expected low-energy satellite of the $(A_1)_2$ band. The equatorially substituted species is of symmetry C_s and the secular equations are

$$A' : \begin{vmatrix} k(2) + t - K & \sqrt{2}c & \sqrt{2}d & \sqrt{2}Rc \\ \sqrt{2}c & k(2) - K & d & Rt \\ \sqrt{2}d & d & k(1) - K & Rd \\ \sqrt{2}Rc & Rt & Rd & R^2 k(2) - K \end{vmatrix} = 0$$

$$A'' : K = k - t. \tag{3.107}$$

The A'' mode correlates with one of the E modes (E_y, say) of the parent species. Of the four A modes, that at lowest frequency is the satellite of the parent E band. The positions of the other bands depend on the relative values of the parameters. Since the four A' modes of the labelled species correlate with $(A_1)_1 + (A_1)_2 + B_2 + E$ of the parent, it is possible to find the position of the B_2 band from the isotope bands and Eq. (3.20). More often, the frequency of the B_2 mode will be known, but Eq. (3.20) will still be of value in verifying the assignments of the isotope bands.

Case (iv): it is certainly true for very many molecules expected to exemplify case (iii) that the B_2 mode is observable in the infrared spectrum. The ideal symmetry of the $M(CO)_5$ grouping must therefore be subject to some kind of a perturbation. Three possibilities suggest themselves.

In many cases, the grouping L in $M(CO)_5L$ is polyatomic and non-linear. Then unless L possesses a four-fold axis of rotation (a condition met in practice by linear ligands but few others) the maximum possible molecular symmetry is C_s, and the "B_2" mode becomes allowed. The B_2 mode is, however, observed even in the spectrum of such species as $Mn(CO)_5X$, where X is a halogen atom. It has been suggested that the observed activity is connected with thermal excitation of bending modes [46], but there is no evidence for the temperature variation of intensity that would then be expected [47]. The most likely cause of the activity in the absence of non-linear ligands is [48] steric congestion, leading to a distortion of the four equatorial CO groups away from four-fold rotational symmetry and towards a D_{2d}-type structure, as in Case (vi) of $M(CO)_4$.

Ligand asymmetry will cause differences in the stretching parameters of the equatorial carbonyls, and hence first order splitting of the E modes. All possible rotamers, however, will be present in solution, so that the observed spectrum will show broadening, rather that incipient resolution.† It seems possible, though this suggestion has yet to be tested, that there will be a relationship between the relative intensity of the infrared B band and the broadening of the E band. Distortion of alternate equatorial CO groups towards and away from the axial groups leads to the existence of two separate equatorial stretching parameters, and hence, in principle, to a splitting of the E band. Such splitting has, however, not so far been reported. The relative intensity of the B_2 mode is given by

$$I(B_2)/I(E) = \tan^2 \omega \qquad (3.108)$$

where ω is the angle through which alternate CO groups are distorted from the equatorial plane.

---

† Rotations in solution may be treated classically for present purposes, since they are slow on the infrared timescale. There is a contrast here between vibrational and n.m.r. spectroscopy, based on the different energies involved. To cause averaging of two infrared signals separated by 1 cm^{-1} would require a process that occurred on average more than 3×10^{10} times in a second.

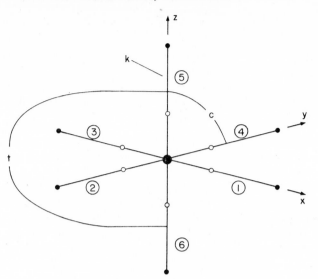

Fɪɢ. 3.21. Force field of octahedral $M(CO)_6$.

$M(CO)_6$

Octahedral $M(CO)_6$ is exemplified by the carbonyls of the metals of the chromium group. The system is of the first class, and the treatment is correspondingly simple. The secular equations are given, in the notation of Fig. 3.21, as

$$A_{1g} : K = k + t + 4c$$
$$E_g \ \ : K = k + t - 2c \qquad (3.109)$$
$$T_{1u} : K = k - t.$$

The symmetry coordinates are, as shown in Fig. 3.22,

$$S(A_{1g}) = [\Delta(1) + \Delta(2) + \Delta(3) + \Delta(4) + \Delta(5) + \Delta(6)]/\sqrt{6}$$
$$S_1(E_g) = [\Delta(1) - \Delta(2) + \Delta(3) - \Delta(4)]/2$$
$$S_2(E_g) = [-\Delta(1) - \Delta(2) - \Delta(3) - \Delta(4) + 2\Delta(5) + 2\Delta(6)]/\sqrt{12}$$
$$S(T_{1u})_x = [\Delta(1) - \Delta(3)]/\sqrt{2}$$
$$S(T_{1u})_y = [\Delta(2) - \Delta(4)]/\sqrt{2}$$
$$S(T_{1u})_z = [\Delta(5) - \Delta(6)]/\sqrt{2}. \qquad (3.110)$$

The ordering

$$K(A_{1g}) > K(E_g) > K(T_{1u})$$

is unambiguous. The A_{1g} and E_g modes are active in the Raman spectrum only, while the T_{1u} mode is active only in the infrared.

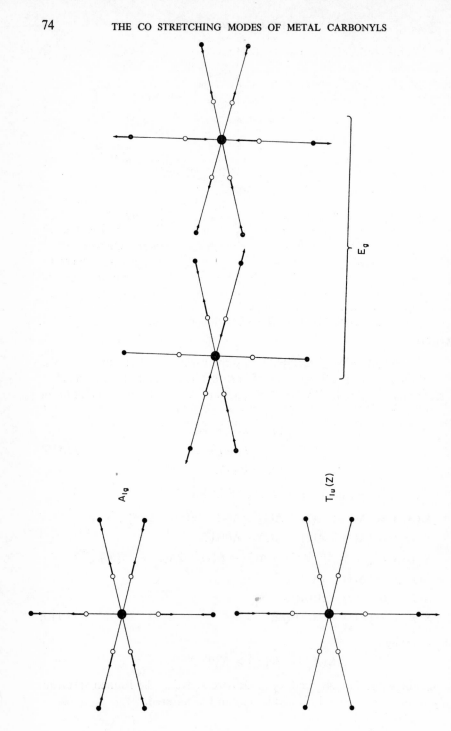

FIG. 3.22. Normal CO stretching modes of octahedral $M(CO)_6$.

The monosubstituted species $M(CO)_5$ $(CO)^*$ occurs at natural abundance to the extent of nearly 7%. The species belongs to the point-group C_{4v}, and the B_2 and E modes correlate with the E_g and T_{1u} modes respectively of $M(CO)_6$. The remaining modes, of symmetry A_1, are given by the secular equation

$$A_1 : \begin{vmatrix} k + t + 2c - K & 2c & 2Rc \\ 2c & k - K & Rt \\ 2Rc & Rt & R^2k - K \end{vmatrix} = 0 \qquad (3.111)$$

Thus, the monosubstituted species should in principle give rise to three isotope bands. One of these is the low energy satellite of the unique infrared-active band, while the other two are at higher energy and weaker.

The effects of small perturbations on the vibrational modes are similar to those of other species showing degenerate modes. If one CO group only, is perturbed relative to the others, so that it acquires a stretching parameter of $k + 5\delta$, while the remaining groups have a parameter $k - \delta$, then the A_{1g} mode is to first order unaffected. The E_g mode is split into modes of effective force constant $k + t - 2c \pm \delta$, while T_{1u} mode is split into one mode of effective force constant $k - t + 2\delta$ and two of effective force constant $k - t - \delta$. Thus it is in principle possible to infer the magnitude and sign of δ for perturbations of this type.

If two mutually cis-CO groups are perturbed, they may be assigned stretching parameters $k + 2\delta$, while the remaining parameters are given the value $k - \delta$. Then the A_{1g} mode is once more unaffected. The E modes are split into modes of effective force constant $k + t - 2c \pm \delta/2$, while the T_{1u} modes are split into two modes of effective force constant $k - t + \delta/2$ and $k - t - \delta$. If three mutually fac-CO groups are perturbed, so that their stretching parameters take the value $k + \delta$, while the remaining parameters take the value $k - \delta$, then none of the modes are affected to first order. This is because, in the normal modes of the unperturbed parent, any distortion of a CO group of parameter $k + \delta$ is accompanied by an equal disturbance of the trans group, the parameter of which is $k - \delta$. Thus a small perturbation of this kind will escape detection.

3.7. The Normal Modes of Polynuclear Carbonyls

There are many compounds containing more than one metal atom coordinated by carbon monoxide. These metal atoms may be connected by metal–metal bonds, or bridging ligands, or both. In any case, the interaction parameters connecting carbonyl groups on the different metal atoms cannot be considered negligible, although they are generally smaller than those between groups attached to the same atom. Since no rigorous studies have as yet been carried out on any polynuclear carbonyl, it is not possible to be precise about the origin of the interaction parameters of the energy factored force field. It is, however, known that these parameters are very sensitive to the relative orientations of the carbonyl groups, and can sometimes be negative. Both these facts seem consistent with a measure of through-space interaction between the MCO units, rather than a mechanism involving orbital following.

The energy factored force field of a polynuclear carbonyl $[M(CO)_n]_x$ may be formulated in two stages. Firstly, symmetry coordinates are set up for the individual $M(CO)_n$ units using their local symmetry. Secondly, these coordinates are combined over the x separate units using the symmetry of the entire molecule. Secular equations may then be written using the symmetry coordinates so generated. In this section, such a treatment is applied to a few simple or commonly met situations. It is possible for metal carbonyl fragments of different idealised local symmetry to be linked in the same molecule; the overall symmetry is then generally low, and it is difficult to draw any detailed inferences from the resulting complex spectrum.

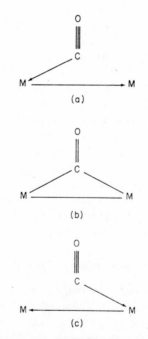

FIG. 3.23. Valence bond representation of bridging CO. In asymmetrically bridging groups, structures (a) and (c) contribute to different extents.

Bridging Carbonyls

Among commonly met bridging ligands is carbon monoxide itself. This is generally coordinated to both metal atoms through the carbon, either symmetrically or asymmetrically. Valence bond representations of both these situations are shown in Fig. 3.23, although an equivalent orbital description is to be preferred [1]. Many compounds are also known in which the carbon atom of a bridging carbonyl group is attached to three metal atoms. Some cases are also known in which the oxygen atom of a carbonyl ligand is coordinated to aluminum [49] or to europium (III) [50], and there seems no reason why such donation (Fig. 3.24) should not also occur to outer transition metals, at least in reaction intermediates or matrix-isolated species.

$$M \longleftarrow \overset{-}{C} \equiv\equiv\equiv \overset{+}{O} \longrightarrow M'$$

FIG. 3.24. CO coordinated through both carbon and oxygen.

The interaction between bridging and terminal carbonyl groups is generally ignored. It is certainly true that cross-terms between bridging and terminal carbonyls occur in the anharmonic correction to the quadratic energy factored force field, since combinations involving both kinds of motion are observable in the spectrum of $Co_2(CO)_8$ in the $4,000 \text{ cm}^{-1}$ region [51], and it seems reasonable to infer that there are cross-terms in the quadratic energy factored force field itself. Nonetheless, the effect of such coupling will be minimised by the separation between the stretching parameters of the two kinds of group. The group frequencies of doubly bridging carbonyls are commonly observed around $1800-1850 \text{ cm}^{-1}$ (for neutral species), while triply bridging CO absorbs at around 1750 cm^{-1} or less and CO bridged through both atoms absorbs at around 1650 cm^{-1}. All these vibrations give rise to bands that are moderately intense in the infrared spectrum.†

The vibrations of bridging carbonyls have not been investigated as thoroughly as those of terminal groups, for many reasons. Some polynuclear species are insoluble or unstable in solution, or are too deeply coloured or photosensitive for Raman studies. The structures of polynuclear carbonyls are sensitive to small changes in chemical composition, and are often of quite low symmetry, so that interpretation of spectroscopic data may be difficult, and it is not easy to prepare series of closely related isostructural species for

---

† It is possible that the absorption due to the vibration of a CO group attached at each end to a transition metal would be moderately weak, since lengthening of the CO bond would cause *opposed* electron motions from both metal atoms into the bridging group.

detailed study, or to be sure that one has done so. Finally, the synthetic chemistry of polynuclear carbonyls and their derivatives is less straightforward than that of mononuclear carbonyls, since the degree of aggregation may change during the course of a reaction and quite complicated equilibria can then be set up. Despite these problems, it is to be hoped that comparative studies will eventually be forthcoming, since they could add appreciably to our knowledge of medium-range electronic interactions in polynuclear species.

For the sake of completeness, we list here some of the possible arrangements of equivalent bridging carbonyl groups:

two groups: the theory is the same as for $M(CO)_2$, Case (i). The selection rules depend on whether or not the groups are connected by a centre of symmetry.

three groups: possible arrangements include C_{3v} (as in $Co_4(CO)_{12}$ [52]) and D_{3h} (as in $Fe_2(CO)_9$ [53]). The theory resembles that for $M(CO)_3$ (Case i).

four groups: these are known to occur in a tetrahedral arrangement (as in $(\pi\text{-}C_5H_5)_4\,Fe_4(CO)_4$ [54] and $Rh_6(CO)_{16}$ [55]). The theory resembles that for $M(CO)_4$ (Case i). Another likely arrangement is a compressed tetrahedron (D_{2d}), Case (vi).

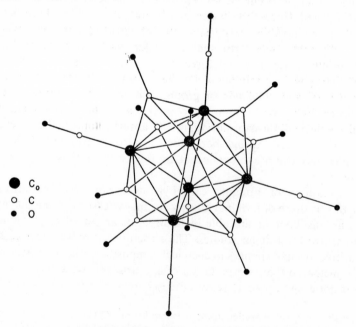

Fig. 3.25. Idealised structure of the anion $Co_6(CO)_{14}{}^{4-}$ [56].

eight groups: these are found in $Co_6(CO)_{14}^{4-}$ [56], (Fig. 3.25) which belongs in the crystal of the tetramethylammonium salt to the point group S_6, and ideally to O_h. In O_h the bridging group vibrations span $A_{1g} + T_{2g}$ (Raman-allowed) + T_{1u} (infrared-allowed) + A_{2u} (forbidden), but solid state and solution ion pair effects may upset the predictions of simple theory, and isomerisation may well be facile.

Terminal Carbonyl Groups

We list here the carbonyl modes of some selected species of the type $[M(CO)_n]_x$; an exhaustive analysis is clearly impracticable. For less symmetric species, the number of parameters is large and the molecular symmetry generally low, so that no detailed analysis of the force field is possible. It is, however, often possible to identify bands due to specific structural units even in such difficult cases, and hence to draw inferences about stereochemistry and bonding.

Rotational isomerism

The barrier to rotation in binuclear systems without bridging groups is not known, but may well be low. The resultant rotation will not cause averaging of interaction constants, since the time-scale of an infrared experiment is too short. A possibly significant effect could be the presence of a range of conformers, covering a continuum of dihedral angles. The effects will be to cause broadening of the observed bands, and possibly also some relaxation of the selection rules that apply to rigorously staggered or eclipsed conformations.

A second kind of rotational isomerism is possible where the metals carry both carbonyl and non-carbonyl ligands, especially if bridging groups are present. In such cases, both trans and gauche, or cisoid and transoid arrangements of terminal ligands can occur, as shown in Fig. 3.26, 3.27.

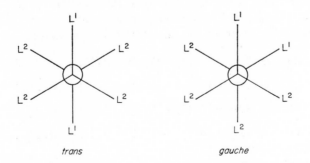

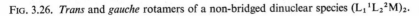

FIG. 3.26. *Trans* and *gauche* rotamers of a non-bridged dinuclear species $(L_1{}^1L_2{}^2M)_2$.

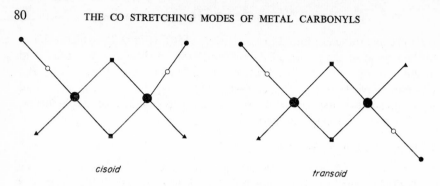

cisoid   transoid

FIG. 3.27. *Cisoid* and *transoid* isomers of a bridged dinuclear species $(LMCO)_2(B)_2$.

These isomers will belong to different point groups and have different spectra, although accidental overlap of bands is a strong possibility. Interconversion of isomers may be too fast in solution for either spectrum to be observed alone, but this will not lead to broadening of the bands. A probable mechanism for interconversion of isomers in bridged species is breaking of the bridges, and it may or may not be possible in such cases to observe bands due to the non-bridged intermediates.

Species $[M(CO)]_x$: If $x = 2$, the theory follows that for $M(CO)_2$, Case (i). The selection rules will depend on whether or not the two CO groups are connected by a centre of symmetry. If they are not, the relative infrared intensities of the two modes will depend on the angle between the oscillating dipoles (which need not now be co-planar), according to Eq. (3.43). The theory for $[M(CO)]_x$ is in general similar to that given for $M(CO)_x$ in the preceding section.

Species $[M(CO)_2]_x$: These may be discussed in terms of the overall coupling of the a' and a'' modes of the individual $M(CO)_2$ units.† Where $x = 2$, the $[M(CO)_2]_2$ unit as a whole may belong to the point-groups D_2, D_{2h}, D_{2d}, C_{2v} or C_{2h} (Fig. 3.28). The overall symmetry coordinates, specified as combinations of the symmetry coordinates of the units, are:

$$S(1) = S(a'(1) + a'(2)) = [\Delta(11) + \Delta(12) + \Delta(21) + \Delta(22)]/2$$

$$S(2) = S(a'(1) - a'(2)) = [\Delta(11) + \Delta(12) - \Delta(21) - \Delta(22)]/2$$

$$S(3) = S(a''(1) + a''(2)) = [\Delta(11) - \Delta(12) + \Delta(21) - \Delta(22)]/2$$

$$S(4) = S(a''(1) - a''(2)) = [\Delta(11) - \Delta(12) - \Delta(21) + \Delta(22)]/2 \qquad (3.112)$$

---

† We assign these in a possibly idealised local symmetry, in this case C_s, and use lower case letters for the symmetry species of the individual units.

and the secular equations are

$$S(1): K = k + i + p + q$$
$$S(2): K = k + i - p - q$$
$$S(3): K = k - i + p - q$$
$$S(4): K = k - i - p + q \tag{3.113}$$

(for D_{2d} only, $p = q$)

The symmetry classifications, and Raman and infrared allowedness, of these symmetry coordinates in the three point-groups considered are

	D_2	D_{2h}	D_{2d}	C_{2v}	C_{2h}
$S(1)$	$A;R$	$A_g;R$	$A_1;R$	$A_1;R,IR$	$A_g;R$
$S(2)$	$B_1;R,IR$	$B_{1u};IR$	$B_2;IR$	$B_1;R,IR$	$B_u;IR$
$S(3)$	$B_3;R,IR$	$B_{3u};IR$	$E;R,IR$	$B_2;R,IR$	$A_u;IR$
$S(4)$	$B_2;R,IR$	$B_{2g};R$		$A_2;R$	$B_g;R$

$$\tag{3.114}$$

Thus at least one mode is infrared inactive. If i is greater than p or q, the highest frequency infrared active band corresponds (except in C_{2v}) to $S(2)$. The ordering of $S(3)$ and $S(4)$ depends on the conventional assignment of labels. The predicted relative intensities of the three active modes in D_2 are

$$I(S(2)) : I(S(3)) : I(S(4)) :: \cos^2 \tfrac{1}{2}\theta(i) : \sin^2 \tfrac{1}{2}\theta(i)\cos^2 \tfrac{1}{2}\theta(p) :$$
$$\sin^2 \tfrac{1}{2}\theta(i)\sin^2 \tfrac{1}{2}\theta(p). \tag{3.115}$$

The D_{2h} case corresponds to $\theta(p) = 0$ so that the intensity of the $S(4)$ mode vanishes. The D_{2d} case corresponds to $\theta(p) = 90°$, so that the intensities of the equivalent $S(3)$ and $S(4)$ modes must be added to give the observed intensity of the E mode.

If Raman or combination data are unobtainable, the force field can only be determined using data for isotopically labelled molecules. If we choose $CO(22)$ to be the labelled group, and take $a'(1)$, $a''(1)$, $\Delta(21)$ and $\Delta(22)$ as a basis set, the secular equation is

$$\begin{vmatrix} k + i - K & 0 & \sqrt{2}(p + q) & \sqrt{2}R(p + q) \\ 0 & k - i - K & \sqrt{2}(p - q) & -\sqrt{2}R(p - q) \\ \sqrt{2}(p + q) & \sqrt{2}(p - q) & k - K & Ri \\ \sqrt{2}R(p + q) & -\sqrt{2}R(p - q) & Ri & R^2k - K \end{vmatrix} = 0 \tag{3.116}$$

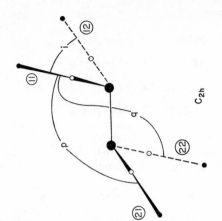

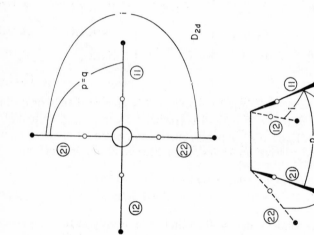

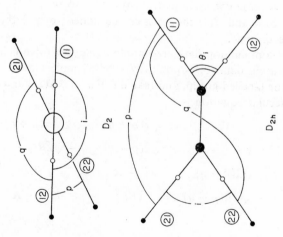

FIG. 3.28. Possible arrangements for species [M(CO)₂]₂ (all CO groups equivalent).

There are, in principle, four bands in the infrared spectrum of the monosubstituted species, though it seems likely that only one of these, the low energy satellite band, will be observable at natural abundance. The force field may, for all cases except C_{2h}, be fitted to this band, but this is, as discussed for $M(CO)_2$, Case (i), a doubtful procedure. Alternatively, q, for instance, may be chosen as an independent variable. The dependence of the other parameters on q is then immediately apparent from Eq. (3.113); and, except perhaps for the D_{2h} and C_{2v} cases, it seems reasonable to confine attention to the range of results for which q is small.

The range of results possible for $[M(CO)_2]_2$ species illustrates the ambiguities that arise with polynuclear species. If only two infrared bands are observed, this may be for one of several reasons. The molecule may belong to one of those point-groups (D_{2h}, D_{2d}, C_{2h}) for which only two bands are expected, or $p - q$ may be very small (so that v_3 and v_4 are accidentally degenerate). Less probably, p and i may be almost equal, so that v_2 and v_3 are accidentally degenerate. Again, if three bands are observed, the molecule could still be assigned either to D_2 or to C_{2v}. In the former case, the missing mode is that at highest frequency, while in the latter, it is probably at lowest, and the different assignments of molecular symmetry will lead to totally different estimates of the force field. Finally, the treatment of $[M(CO)_2]_2$ given here has assumed that the carbonyl groups are connected by symmetry elements. This assumption need not be valid for a particular molecule, in which case four bands should in principle be observable; but there may well be accidental degeneracy or lack of intensity, suggesting a higher degree of symmetry than is actually present.

Species $[M(CO)_3]_x : x = 2$. This case is exemplified by such species as $[LCo(CO)_3]_2$ [57] and $Fe_2(CO)_9$ [53]. In addition, $[\pi\text{-}C_5H_5Mo(CO)_3]_2$ [58, 59] is an intensity example of a molecule whose modes may apparently be analysed using an idealised local symmetry for each moiety higher than is in fact present.

The symmetry coordinates for each moiety are as given in the treatment of $M(CO)_3$, Case (i), in the previous section. The overall symmetry coordinates, secular equations and selection rules are as follows:

D_{3d} *(Figure 3-29a)*

$A_{1g}(R)$: $S = [\Delta(11) + \Delta(12) + \Delta(13) + \Delta(21) + \Delta(22) + \Delta(23)]/\sqrt{6}$

$\qquad\quad K = k + 2i + 2p + q$

$A_{2u}(IR)$: $S = [\Delta(11) + \Delta(12) + \Delta(13) - \Delta(21) - \Delta(22) - \Delta(23)]/\sqrt{6}$

$\qquad\quad K = k + 2i - 2p - q$

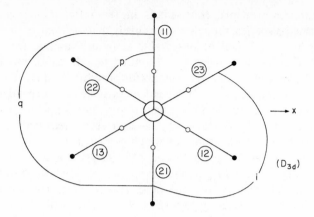

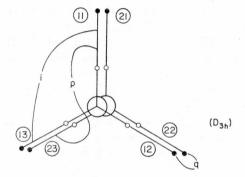

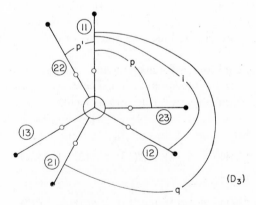

FIG. 3.29. Possible arrangements for species $[M(CO)_3]_2$ (all CO groups equivalent). (a) D_{3d} (b) D_{3h} (c) D_3.

$E_g(R)$: $S_{xz} = [\Delta(12) - \Delta(13) + \Delta(22) - \Delta(23)]/2$

$S_{yz} = [2\Delta(11) - \Delta(12) - \Delta(13) + 2\Delta(21) - \Delta(22) - \Delta(23)]/\sqrt{12}$

$K = k - i - p + q$

$E_u(IR)$: $S_x = [\Delta(12) - \Delta(13) - \Delta(22) + \Delta(23)]/2$

$S_y = [2\Delta(11) - \Delta(12) - \Delta(13) - 2\Delta(21) + \Delta(22) + \Delta(23)]/\sqrt{12}$

$K = k - i + p - q$ (3.117)

D_{3h} (Fig. 3-29b)

$A_1'(R)$, A_2'' (IR), E' (R,IR) and E'' (R) correspond in form to A_{1g}, A_{2u}, E_g and E_u, respectively, of the D_{3d} case above.

D_3 (Fig. 3-29c)

$A_1(R)$ and $A_2(IR)$ correspond in form to A_{1g} and A_{2u} of D_{3d}, and the secular equations can be generated from those for D_{3d} by replacing $2p$ with $p + p'$. The forms of the E modes will depend on the relative values of p and p', so that generalisation is not possible. The two E modes will not be distinguished by symmetry type and will formally be allowed to show both infrared and Raman activity.

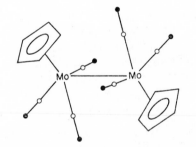

FIG. 3.30. Structure of $[\pi\text{-}C_5H_5Mo(CO)_3]_2$ in the solid [58].

The vibrations of $[(\pi\text{-}C_5H_5) Mo(CO)_3]_2$, which in the solid has the centrosymmetric structure of Fig. 3.30 [58], may be regarded as a superposition of those of interacting fragments $[M(CO)]_2$ and $[M(CO)_2]_2$, of symmetry C_{2h}. Of the expected three infrared peaks for such a system, only two are observed in CCl_4, but that at lower energy is broad and in hydrocarbons is resolvable into two separate peaks. Thus the infrared spectrum in CCl_4 is deceptively simple, resembling that of a species of symmetry C_{3v}. It is noteworthy that the molecule shows an extra, higher energy band in methylene chloride or in tetrahydrofuran. This band is not observable in CCl_4 and is weak in hydrocarbons; it may be assigned to non-centrosymmetric rotamers. Coupling across the metal–metal bond is not negligible, and polar solvents presumably stabilise non-centrosymmetric conformers. The higher energy band can be detected even in non-polar solvents by combination spectroscopy, showing that the polar solvent affects the symmetry rather than the force field.

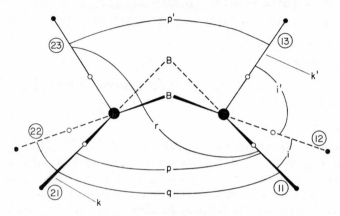

FIG. 3.31. Structure and force field of species $M_2(CO)_6$ (bridge)$_2$.

A closely related problem is posed by the vibrations of a large number of species (e.g. $Fe_2(CO)_6(SR)_2$ [60], $Co_2(CO)_6$ (acetylene) [61], the bridged form of $Co_2(CO)_8$ [62]) of symmetry C_{2v} (Fig. 3.31). The symmetry coordinates and the corresponding secular equations are

$$S(A_1)_1 = [\Delta(11) + \Delta(12) + \Delta(21) + \Delta(22)]/2$$
$$S(A_1)_2 = [\Delta(13) + \Delta(23)]/\sqrt{2}$$
$$S(B_1)_1 = [\Delta(11) + \Delta(12) - \Delta(21) - \Delta(22)]/2$$
$$S(B_1)_1 = [\Delta(13) - \Delta(23)]/\sqrt{2}$$
$$S(B_2) = [\Delta(11) - \Delta(12) + \Delta(21) - \Delta(22)]/2$$
$$S(A_2) = [\Delta(11) - \Delta(12) - \Delta(21) + \Delta(22)]/2 \quad (3.119)$$

and

$$A_1 : \begin{vmatrix} k + i + p + q - K & \sqrt{2}(i' + r) \\ \sqrt{2}(i' + r) & k' + p' - K \end{vmatrix} = 0$$

$$B_1 : \begin{vmatrix} k + i - p - q - K & \sqrt{2}(i' - r) \\ \sqrt{2}(i' - r) & k' - p' - K \end{vmatrix} = 0$$

$$B_2 : K = k - i + p - q$$
$$A_2 : K = k - i - p + q. \quad (3.120)$$

Of the six modes, five are infrared-active, while A_2 is inactive. All the modes, however, are active in the Raman spectrum. There are a total of eight parameters, so that even if all the frequencies are known the force field is underdetermined, with two degrees of freedom. Thus the range of possible solutions is too great even to be displayed by the independent variable method. A force field may be specified only by the imposition of arbitrary approximations, or by the use of data for isotopically labelled species.

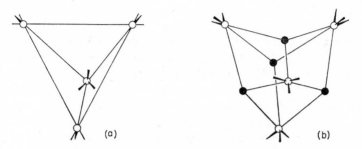

FIG. 3.32. Structures of $Ir_4(CO)_{12}$ [63] and $[Mn(CO)_3SEt]_4$ [64].

If $x = 4$, the molecule belongs to the point-group T_d. Two structures are possible, according to whether the carbonyls are staggered or eclipsed with the metal–metal directions (Fig. 3.32). The former situation arises in $Ir_4(CO)_{12}$ [63], where the structure is held together by metal–metal bonds. The latter is exemplified by compounds of the type $[Mn(CO)_3SR]_4$, where the tetrahedron is held together by face-bridging ligands [64].

The secular equations are

$$A_1 : K = k + 2i + 2a + b + 2c + 4d$$

$$E \ : K = k - i - a + b + 2c - 2d$$

$$T_1 : K = k - i - a - b + 2d$$

$$T_2 : \begin{vmatrix} P - K & Q \\ Q & R - K \end{vmatrix} = 0 \qquad (3.121\text{a})$$

where

$$P = k + 2i - 2a/3 - b/3 - 2c/3 - 4d/3$$

$$Q = 2(2a - 2b + 2c - 2d)/3$$

$$R = k - i + 5a/3 + b/3 - 4c/3 - 2d/3 \qquad (3.121\text{b})$$

and in the notation of [31]

$$k = k(1, 1)$$
$$i = k(1, 2)$$
$$a = k(1, 6)$$
$$b = k(1, 12)$$
$$c = k(1, 4)$$
$$d = k(1, 5). \tag{3.122}$$

The force field is undetermined although only one kind of CO group is present. This is because interaction is possible between the a_1 mode of one $M(CO)_3$ fragment and the e modes of its neighbours. There is reason to believe that the particular solution $d = 0$ is acceptable for $[Mn(CO)_3SR]_4$, in which there is no metal–metal bond and the groups $CO(1)$, $CO(5)$ are distant. This solution is, however, probably unacceptable for $Ir_4(CO)_{12}$ [31]. $[M(CO)_4(C_{2v})]_x$: $x = 2$ *or* 3. The case where $x = 2$ is exemplified by a wide range of species of the general type $[M(CO)_4]_2$ [bridge]$_2$, with or without metal–metal bonding, while $x = 3$ is exemplified to date only by bonded or semi-bonded species such as $Os_3(CO)_{12}$, $Mn_3(CO)_{12} H_3$ [65]. The symmetry coordinates may be written as for two interacting systems $[M(CO)_2]_x$,

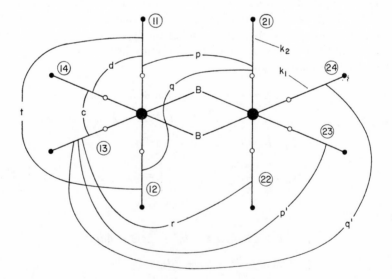

FIG. 3.33. Force field of species $[M(CO)_4]_2$[bridge]$_2$.

representing the axial and equatorial carbonyl groups respectively. Thus for $x = 2$, we have (Fig. 3.33)

$$\begin{aligned}
S(A_g)_a &= [\Delta(11) + \Delta(12) + \Delta(21) + \Delta(22)]/2 &\quad (R)\\
S(A_g)_e &= [\Delta(13) + \Delta(14) + \Delta(23) + \Delta(24)]/2 &\quad (R)\\
S(B_{1u})_a &= [\Delta(11) + \Delta(12) - \Delta(21) - \Delta(22)]/2 &\quad (IR)\\
S(B_{1u})_e &= [\Delta(13) + \Delta(14) - \Delta(23) - \Delta(24)]/2 &\quad (IR)\\
S(B_{2u}) &= [\Delta(11) - \Delta(12) + \Delta(21) - \Delta(22)]/2 &\quad (IR)\\
S(B_{3u}) &= [\Delta(13) - \Delta(14) + \Delta(23) - \Delta(24)]/2 &\quad (IR)\\
S(B_{2g}) &= [\Delta(13) - \Delta(14) - \Delta(23) + \Delta(24)]/2 &\quad (R)\\
S(B_{3g}) &= [\Delta(11) - \Delta(12) - \Delta(21) + \Delta(22)]/2 &\quad (R) \quad (3.123)
\end{aligned}$$

giving secular equations

$$A_g : \begin{vmatrix} k(2) + t + p + q - K & 2d + 2r \\ 2d + 2r & k(1) + c + p' + q' - K \end{vmatrix} = 0$$

$$B_{1u} : \begin{vmatrix} k(2) + t - p - q - K & 2d - 2r \\ 2d - 2r & k(1) + c - p' - q' - K \end{vmatrix} = 0$$

$$\begin{aligned}
B_{2u} &: K = k(2) - t + p - q\\
B_{3u} &: K = k(1) - c + p' - q'\\
B_{2g} &: K = k(1) - c - p' + q'\\
B_{3g} &: K = k(2) - t - p + q. \quad (3.124)
\end{aligned}$$

The force field is underdetermined, with two degrees of freedom. Exact studies would require the use of data from isotopically labelled species, involving difficulties in assignment. The separation between the highest A_g and B_{1u} modes has, however, been proposed as a measure of the mutual coupling of the two moieties.

The theory for $[M(CO)_4]_3$ (Fig. 3.34) is rather similar; instead of in-phase

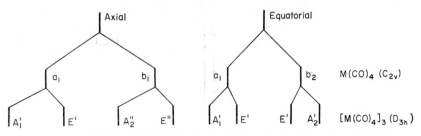

FIG. 3.34. Splitting of CO Modes in $Os_3(CO)_{12}$ (final order not predictable).

and out-of-phase combinations over the two metal centres, the combinations are chosen so as to be of symmetry A or E over the three centres. The force field has no fewer than four degrees of freedom, and there are eight vibrational modes, of which only four are infrared active.

$[M(CO)_4 (C_{4v})]_2$: The theory for the coupling of two $M(CO)_4$ units of symmetry C_{4v} closely resembles that for the coupling of two $M(CO)_3$ units of symmetry C_{3v}. The $M(CO)_4$ moieties may be staggered, eclipsed or intermediate, belonging to the point-groups D_{4d}, D_{4h} and D_4 respectively (Figs. 3.35(a), (b), (c)). The normal coordinates may be constructed from

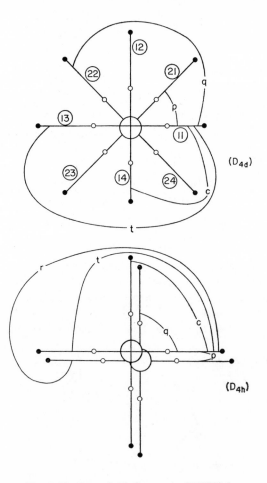

FIG. 3.35. Force fields for species $[M(CO)_4]_2$.

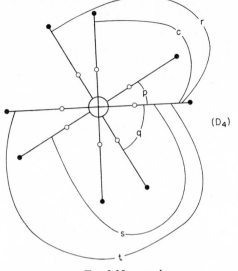

(D_4)

FIG. 3.35—*contd.*

linear combinations of the local a_1, b_2, e fragments. The combinations of the non-degenerate local coordinates may be written in all cases as

$$S(1) = S(a_1(1) + a_1(2)) = [\Delta(11) + \Delta(12) + \Delta(13) + \Delta(14) + \Delta(21) + \Delta(22) + \Delta(23) + \Delta(24)]/2\sqrt{2}$$

$$S(2) = S(a_1(1) - a_1(2)) = [\Delta(11) + \Delta(12) + \Delta(13) + \Delta(14) - \Delta(21) - \Delta(22) - \Delta(23) - \Delta(24)]/2\sqrt{2}$$

$$S(3) = S(b_2(1) + b_2(2)) = [\Delta(11) - \Delta(12) + \Delta(13) - \Delta(14) + \Delta(21) - \Delta(22) + \Delta(23) - \Delta(24)]/2\sqrt{2}$$

$$S(4) = S(b_2(1) - b_2(2)) = [\Delta(11) - \Delta(12) + \Delta(13) - \Delta(14) - \Delta(21) + \Delta(22) - \Delta(23) + \Delta(24)]/2\sqrt{2} \qquad (3.125)$$

where the labelling and activities of these symmetry coordinates are

	D_{4h}	D_{4d}	D_4
$S(1)$	A_1; R	A_{1g}; R	A_1; R
$S(2)$	B_2; IR	A_{2u}; IR	A_2; IR
$S(3)$		B_{2g}; R	B_2; R'
$S(4)$	E_2; R	B_{1u}; inactive	B_1; R $\qquad (3.126)$

In D_{4d}, the local e symmetry coordinates may be combined to give overall coordinates E_1 (infrared-active, Raman-inactive) or E_3 (infrared-active, Raman-active):

$$S(E_1)_x = S(e_x(1) + e_x(2)) = [\Delta(11) - \Delta(13)]/2 + [\Delta(21) - \Delta(22)$$
$$- \Delta(23) + \Delta(24)]/2\sqrt{2}$$
$$S(E_3)_x = S(e_x(1) - e_x(2)) = [\Delta(11) - \Delta(13)]/2 - [\Delta(21) - \Delta(22)$$
$$- \Delta(23) + \Delta(24)]/2\sqrt{2} \tag{3.127}$$

The secular equations are

$$A_1 : K = k + 2c + t + 2p + 2q$$
$$B_2 : K = k + 2c + t - 2p - 2q$$
$$E_1 : K = k - t + \sqrt{2}(p - q)$$
$$E_2 : K = k - 2c + t$$
$$E_3 : K = k - t + \sqrt{2}(q - p) \tag{3.128}$$

and the problem is determinate provided only all five bands can be assigned.

In D_{4h}, the local e symmetry coordinates may be combined to give

$$S(E_u)_x = S(e_x(1) + e_x(2)) = [\Delta(11) - \Delta(13) + \Delta(21) - \Delta(23)]/2$$
$$S(E_g)_x = S(e_x(1) - e_x(2)) = [\Delta(11) - \Delta(13) - \Delta(21) + \Delta(23)]/2. \tag{3.129}$$

The secular equations are

$$A_{1g} : K = k + 2c + t + p + 2q + r$$
$$A_{2u} : K = k + 2c + t - p - 2q - r$$
$$B_{2g} : K = k - 2c + t + p - 2q + r$$
$$B_{1u} : K = k - 2c + t - p + 2q - r$$
$$E_g \;\; : K = k - t - p + r$$
$$E_u \;\; : K = k - t + p - r. \tag{3.130}$$

In D_4, it is not possible to specify *a priori* unique E modes, since the form of the molecular vibrations will depend on the relative values of p, q, r and s. In D_{4d} and D_{4h}, the number of parameters is equal to the number of bands so that if the experimental problem of observing and assigning all the frequencies can be solved, the force field is uniquely specified.

$[M(CO)_5(C_{4v})]_2$: Closely related to the previous cases is that of two coupled $M(CO)_5$ units (Fig. 3.36). The only modes of the $[M(CO)_4]_2$ fragment to be affected are A_1 and B_2; for these the secular determinants in D_{4d} become

$$A : \begin{vmatrix} k(a) + j - K & 2d + 2h \\ 2d + 2h & k(e) + 2c + t + 2p + 2q - K \end{vmatrix} = 0$$

$$B : \begin{vmatrix} k(a) - j - K & 2d - 2h \\ 2d - 2h & k(e) + 2c + t - 2p - 2q - K \end{vmatrix} = 0 \qquad (3.131)$$

and the force field is again underdetermined, with two degrees of freedom.

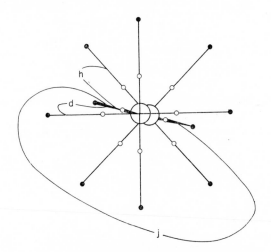

FIG. 3.36. Force field for species $[M(CO)_5]_2$ (staggered).

Intensity Ratios in Polynuclear Carbonyls

It was stated in Section (3.5) above that the direction of the oscillating dipole associated with a carbonyl group need not coincide with that of the group itself. This is particularly true for polynuclear species, where changes in the electron demand of the metal atoms during a vibration may cause orbital following in metal–metal bonds. Such orbital following has been invoked to explain the high infrared intensity of the upper B_1 mode of $Mn_2(CO)_{10}$ [66]. This mechanism would also be expected to enhance the intensities of the B_2 modes of species $[LM(CO)_4]_2$ (D_{4d}) and $[LM(CO)_3]_2$ (D_{3d}), but in species of type $[R_3PMn(CO)_4]_2$ and $[R_3PCo(CO_3)]_3$, the B_2 bands are extremely weak [67, 68]. At the time of writing the reasons for these varying effects are not fully understood.

Orbital following in metal–metal bonds has also been invoked for species of type $[M(CO)_4]_x$ [69]. Thus the highest B_{1u} modes for species of type $Mn_2(CO)_8$ (bridge)$_2$ are relatively weak, that of $[Mn(CO)_4H]_3$ is of intermediate intensity, and those in $Ru_3(CO)_{12}$ and $Os_3(CO)_{12}$ are about as strong as the other allowed bands. In this case at least, metal–metal bonding appears to enhance intensity, but whether this is because of orbital following in the metal–metal bonds themselves, or because of coupling across these bonds of axial and equatorial carbonyl groups, it is impossible as yet to say.

Chapter 4

Experimental Methods in Metal Carbonyl Vibrational Spectroscopy

This Chapter surveys the experimental technique for obtaining meaningful spectra of metal carbonyls. The emphasis is on the $v(CO)$ region throughout, but the other spectral regions are also discussed briefly. Because the spectra of metal carbonyls are informative, attention to the details of technique is important, and this is as true for the routine "fingerprinting" use of spectroscopy as it is in more refined studies.

4.1. Principles of Infrared Experimentation

The Design of Infrared Spectrometers

A double-beam recording spectrometer (Fig. 4.1) comprises a source light from which is reflected by concave mirrors M so as to pass in more or less parallel beams through sample and reference chambers, a rotating beam-selector mirror R, a monochromator, and a detector. The path of the light through the spectrometer is defined by slits S. If the amounts of light passing through sample and reference chambers are unequal, the detector signal will vary with a frequency equal to that of the rotation of the mirror R (typically around 10 revolutions per second). Accordingly, the varying component of the detector signal is amplified and used to alter the position of the attenuator A until sample and reference signals become equal. Additional attenuators A', A'' may be present; these are fixed by the operator before running a spectrum to correct the 100% and 0% points of the transmission scale. The motion of the attenuator A is coupled to that of the recorder pen. As the spectrum is scanned, the monochromator is re-set continuously, so that the wavenumber selected varies linearly with time.† The re-setting of the

† Some instruments produce spectra linear in wavelength rather than wavenumber. Wavelength (microns) $\times$ wavenumber $(cm^{-1}) = 10,000$.

monochromator is synchronous with the movement of the chart paper, and so the pen trace gives a record of the spectrum. The chart paper may be pre-calibrated, or the instrument may incorporate wavenumber marker pens.

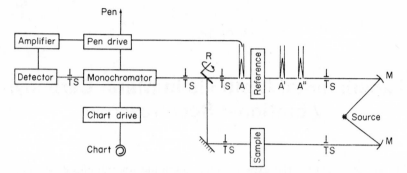

Fig. 4.1. Typical schematic layout of a modern infrared spectrometer.

The source is simply an electric resistor, heated by the passage of a current, and operating as a hot body. The working temperature is a compromise between two requirements. The higher the temperature, the greater the source output, but high source temperatures also favour high rather than low frequencies. This increases the very wide range of source power with which the spectrometer must be able to function. The monochromator may be either a grating or a prism. In the former case the frequency is scanned by altering the angle between the grating and the incident light, while in the latter, the beam is reflected back through the prism to the detector and the frequency is selected by altering the angle of the reflecting mirror. The detector is a heat measuring device, since the energies to be studied are too low to activate electronic processes. Devices that have been used include thermocouples, bolometers and Golay detectors. A thermocouple generates a voltage when heated, while a bolometer undergoes a change in resistance. A Golay detector is a gas-filled chamber with a thin, light reflecting wall. As this is heated, the gas pressure increases, causing distortion of this wall and a shift in the position of an image reflected from it.

The strength of the detector signal depends, among other things, on the width of the slits. This slit-width is generally made to vary automatically during scanning, according to a pre-determined program, so as to compensate for the fall-off of source power at low wave-numbers. An alternative procedure used in some spectrometers is to monitor the detector energy and cause the slits to vary so as to keep this constant. This makes the quantitative study of absorption intensities more difficult, and causes even greater difficulties in the quantitative analysis of mixtures, as discussed in

Section (4.4) below. An alternative to changing the slit-widths is variation of the degree of amplification, or "gain", imposed on the detector signal. The disadvantage of this method is that an increase in gain is always accompanied by a more than proportionate increase in the degree of electrical noise (random fluctuations unrelated to signal) in the amplifier output.

Beer's and Lambert's Laws

The absorption of light passing through a homogeneous sample is first order in the light intensity, so that the intensity of a monochromatic beam decays exponentially in passing through the sample. This is *Lambert's law*, which is obeyed in all cases.

If altering the concentration of a substance has no effect on its light-absorbing power, then the absorption (corrected for light loss by cell and solvent) will be first order in the concentration of that substance. This is *Beer's law*. Beer's law is obeyed by a material only if there are no changes in its degree of aggregation over the concentration range studied, and if the effect of the concentration variation on solvent–solute and solute–solute interactions is too small to affect the spectrum. Whether this is so in any particular case can only be decided by experiment.

Absorbance, Optical Density, and Extinction Coefficient

The ability of a sample to absorb light may be expressed in several different ways. The simplest measure is given by the *transmittance*, or degree of the light incident on a sample that is transmitted; this varies from 100% for a perfectly transparent sample to 0% for one that is perfectly opaque. Most infrared spectrometers plot data in this way. Unfortunately, percentage transmittance is not additive but multiplicative.

A more useful measure of absorbing power is the *absorbance*, which is the negative logarithm of the transmittance:

$$A = - \log_{10} T = - \log_{10} (I/I_0) \qquad (4.1)$$

(here I_0 is the intensity of the incident light, and I that of the light transmitted).

Absorbance is an additive property and is proportional to pathlength through a homogenous sample. The absorbance for unit pathlength is known as the *optical density D*:

$$D = A/l = [- \log_{10} (I/I_0)]/l. \qquad (4.2)$$

For materials that obey Beer's law, optical density is proportionate to concentration. It is then possible to define a *molar extinction coefficient*, ε, such that

$$\varepsilon = D/C \qquad (4.3)$$

(where C is the molar concentration of an absorbing material) and

$$A = \varepsilon C l. \qquad (4.4)$$

If more than one absorbing material is present, the effects of these on the absorbance are additive:

$$A = \sum_a \varepsilon_a C_a l \qquad (4.5)$$

Thus absorbance is a more fundamental quantity than transmittance, and it is unfortunate that many infrared spectrometers cannot present data in this way.

4.2. The Selection of Operating Conditions: Infrared

Choice of Instrument

Few readers will be in the happy position of being able to dedicate an instrument exclusively to the spectroscopy of metal carbonyls. Nonetheless, it is reasonable to expect spectrometers to be available that can be used in metal carbonyl research and the special requirements implied should be born in mind when general purpose purchases are contemplated. This is at least as true for routine or teaching instruments as for more sophisticated and expensive equipment. The freedom of choice open to the operator of a simple instrument is smaller and the possibility of design features proving troublesome is correspondingly increased.

A grating instrument is much to be preferred to one that uses a prism as monochromator. The amount of power of given wavelength transmitted by a prism instrument is less, and so a larger time must be taken to record a spectrum of given quality. Not only is this more tedious, but the possibility of sample decomposition in the beam is thereby increased.

One annoying feature of many instruments is the existence of a mandatory scale change (and sometimes also a change in the monochromator unit) at $2,000\ \mathrm{cm}^{-1}$. It is customary to display the higher energy part of the normal infrared spectrum at lower ordinate expansion than the lower part, and the position chosen for the changeover corresponds to a frequency of limited interest in conventional organic spectroscopy. Unfortunately, the break at $2,000\ \mathrm{cm}^{-1}$ implies a distortion of the overall appearance of the spectrum, and it becomes particularly difficult to detect shoulders or other weak features near the strong bands that often occur around this frequency.

Further complications can arise from differences in instrument error above and below the changeover, and from the possibility of variation in the actual wave-number at the nominal $2,000 \, cm^{-1}$ point (so that a small part of the spectrum is repeated or lost). Changes from one spectrum to another in the size of the above effects can lead to spurious small differences. If the purchase of an instrument with a scale-change in the carbonyl region is contemplated, the manufacturers should be consulted at an early stage; it is in some cases possible to re-locate the scale change above 2100 or below $1800 \, cm^{-1}$, where it will prove less troublesome.

A valuable feature of many instruments is variable chart speed, which functions as a wavenumber scale expansion and improves the readability of the spectrum displayed. The wavelength calibration of the expanded spectrum may however be rather tedious for routine work, especially if the chart paper is designed for use at normal expansion. The author has even found himself dealing with one instrument where in the ordinate expansion mode the chart continued to be fed out during an unavoidable grating change at $2,000 \, cm^{-1}$. A further valuable feature is ordinate expansion, similar to the 0–10% and 90–100% transmittance scales found on many visible-UV instruments. On many infrared instruments, this facility is not available at all, while on some others ordinate expansion may be obtained only by using an external recorder. This is regrettable, particularly since a 90–100% transmittance scale can greatly facilitate the location of weak bands. A linear absorbance scale is a major convenience in reaction monitoring as well as in quantitative studies.

While there may be no firm plans, when an instrument is first purchased, to use the specialised methods of Section 4.4 below, it is obviously desirable to leave options open. Therefore information should be sought about the compatibility of a proposed instrument with polarisers, beam condensers, variable temperature cells, and gas cells. Apparently minor problems of mounting for special purpose cells may lead to important uncertainties in alignment and hence to loss of effective power. As well as problems in mounting, there may be space problems, especially if the sample and reference beams are close together.

Some special applications may impede the use of the full height of the spectrometer beam. In such cases, attenuator design becomes important. A good attenuator will operate uniformly over the entire beam height, so that the intensity of any part of this remains truly representative under all conditions.

The spectrum of water vapour overlaps the metal carbonyl region, causing oscillations in slit-width or gain, and possible spurious bands. This problem can be minimised in some instruments by the passage of a slow stream of dry nitrogen or air.

Finally, the location of an instrument requires some thought. Since temperature affects the optics and hence the wavenumber calibration of instruments, a room of even temperature is obviously desirable, and although some precision instruments have thermostatted optics, direct sunlight on or near an instrument should always be avoided. Damp is undesirable, especially in a prism instrument, for which desiccation is often a necessary part of maintenance. The possible effects of acidic, alkaline or oxidising atmospheres on mirror surfaces should be obvious, and it is thus very important that even routine spectrometers used for simple reaction monitoring should be housed in instrument rooms rather than in the preparative laboratory.

Adjustment of Operating Conditions

The operator generally has some control over instrumental slit width, gain, and recording speed, as well as over the 100% and sometimes the 0% points of the transmission scale.

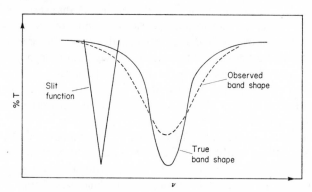

FIG. 4.2. Effect of finite slit width on observed shape of an absorbance band.

Typically, slit-width is adjusted automatically, but a selection of slit programs is available. A narrow slit has the advantage of increasing resolution (up to a limit, imposed by the quality of the monochromator), and is therefore desirable when the sharp, strong bands of metal carbonyls are to be studied. Widening the slit decreases the resolution, and the "optical slit width" is the width of the band of wavenumbers actually sampled. Unless the optical slit width is narrow compared with the true width of an absorption band, spurious broadening will arise (Fig. 4.2). A narrow slit, however, means a relatively weak signal, so that more amplification ("gain") of the detector signal may be necessary before this can be used to drive the attenuator and the pen. Increasing amplification inevitably increases the level of noise in the recorded spectrum. Noise tends, by its very nature, to average out over a period of time, and so, if the spectrum is run more slowly, it becomes easier to distinguish between noise and genuine weak signals. A slower scan, however, increases the possibility of sample decomposition, as well as taking up more operator and instrument time. The choice of the conditions for measuring a particular spectrum must for these reasons depend on the use to which it is to be put.

More routine is adjustment of the 100% and 0% transmittance controls (attenuators A', A'' of Fig. (4.1)). The 100% or "baseline" control is used to balance the intensities of sample and reference beams. This balancing may be carried out with both sample and reference chambers empty, but it is usually better to carry out the adjustment with the sample cell in place and containing (where appropriate) solvent, but no sample. The reference chamber should contain a cell as similar as possible to the sample cell (except for the absence of the sample material itself) which is then left in place while the spectrum is collected.

The baseline need not actually be set at 100% transmittance, and, unless the chart allows for nominal transmittance values greater than 100% some rather lower value is generally chosen. This is because (for reasons discussed more fully in Section 4.3) baselines are often not perfectly horizontal. A baseline set to 100% at one wavenumber may move to 110%

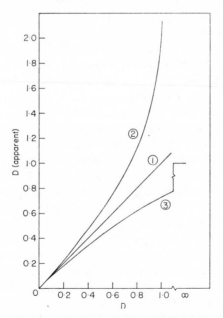

Fig. 4.3. Effect of error in zero transmittance setting. (1) Correct setting. (2) True zero set at -10%. (3) True zero set at 10%.

at a different frequency, so that weak peaks will be lost over part of the spectrum. Repeated use of an instrument with the baseline "off scale" can cause damage to the pen mechanism, and it is important that the baseline is not left in this position between spectra. For routine work, the baseline can be set with the actual sample in place, at a frequency where it does not absorb, but for accurate work the actual form of the baseline throughout the region of interest must be determined.

Although the setting of the baseline is somewhat arbitrary, the correct zero transmission line is uniquely defined. An arbitrary baseline merely shifts all absorbances of the same frequency by an equal amount. An incorrect zero transmission reading affects high more than low absorbances (Fig. (4.3). The zero transmission reading is taken with the reference cell in place and the sample beam completely cut off by a shutter or opaque object. The adjustment may be either optical or electronic, i.e., it may involve altering the setting of an attenuator A'', or changing the gain program.

Wavenumber Calibration

There are a number of reasons why the nominal wavenumber of an absorption should differ from the true value. These include imperfection of the optics (intrinsic, or due to temperature fluctuation), free play in the monochromator, chart drive, or wavenumber marker pen, and incorrect positioning and dimensional instability of precalibrated chart paper. Thus, calibration of instruments is essential, and that of each individual spectrum highly desirable. Ideally, calibration should take place at two wavenumbers, just above and just below the bands of interest, since not only the individual wavenumber setting but the degree of optical dispersion is liable to error. Ideally, also, calibration should take place in a single spectroscopic pass, to avoid errors in re-setting the chart. Where there is a scale change at $2,000\,cm^{-1}$, especially if this is accompanied by change in the monochromator unit, the regions of the spectrum above and below the changeover point should each be separately calibrated.

Reference spectra for the calibration of both high- and low-resolution instruments have been published by the International Union of Pure and Applied Chemistry [70]. The detailed appearance of a calibrant spectrum will, however, be dependent on the spectrometer used and its adjustment. Suitable materials for calibration of accurate instruments in the carbonyl region are CO, DCl and H_2O vapours, and spectra of these are shown in Figs 4.4–4.6. Routine spectra may be calibrated using the absorption lines of polystyrene at 1944 and $1601\,cm^{-1}$, the characteristic appearance of which is shown in Fig. 4.7. The wavenumbers corresponding to some useful peaks are listed in Table 4.1. Suitable calibrants for other regions of the spectrum are described in [70].

TABLE 4.1. Selected Frequencies for CO, DCl and H_2O calibrants (a).

Band	Frequency (b)	Band	Frequency
CO: c, d 1	2161·97	CO: (c, d) 30	2037·03
2	2158·30	31	2032·35
3	2154·60	32	2027·65
4	2150·86	33	2022·91
5	2147·08	34	2018·15
6	2139·43	35	2013·35
7	2135·55	DCl (e, f) 1	2080·28
8	2131·64	2	2069·27
9	2127·69	3	2058·05
10	2123.70	4	2046·61
11	2119·68	5	2034·96
12	2115·63	6	2023·10
13	2111·55	7	2011·04
14	2107·42	8	1998·78
15	2103·27	9	1986·32
16	2099·09	10	1973·66
17	2094·87	11	1960·82
18	2090·61	12	1947·78
19	2086·32	13	1934·57
20	2082·01	14	1921·17
21	2077.65	15	1907·59
22	2073·26	H_2O (g, h) 1	1942·60
23	2068·85	2	1918·05
24	2064·40	3	1895·19
25	2059·91	4	1889·19
26	2055·40	5	1869·35
27	2050·86	6	1825·24
28	2046·28	7	1810·63
29	2041·66	8	1790·9(4)

(a) Data from [70] and from Plyer, E. K., Danti, A., Blaire, L. R., and Tidwell, E. D., *J. Res. Nat. Bureau Standards*, **A64**, 29 (1960).
(b) $\bar{\nu}$ (cm^{-1} in vacuo).
(c) Lines numbered as in Fig. 4.4.
(d) CO bands $1-33$ known to better than 0.01 cm^{-1}.
(e) Lines numbered as in Fig. 4.5.
(f) Data refer to more intense ($D^{35}Cl$) line of each pair.
(g) Lines numbered as in Fig. 4.6.
(h) The appearance of these lines is sensitive to operating conditions, but they are adequate for studies requiring no better than 0.1 cm^{-1}.

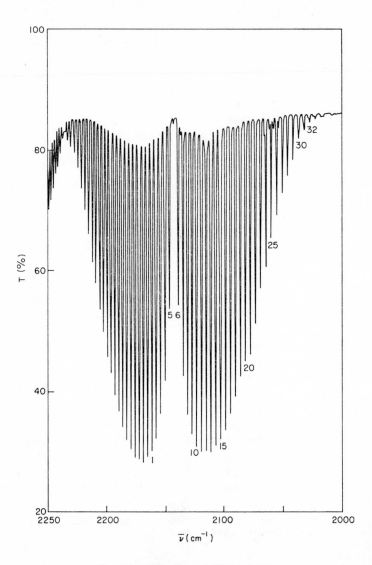

FIG. 4.4. Absorption of CO (coal gas).

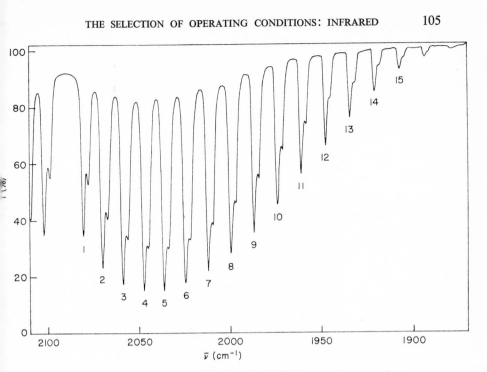

FIG. 4.5. Absorption of DCl (R branch, PE 257 Spectrophotometer).

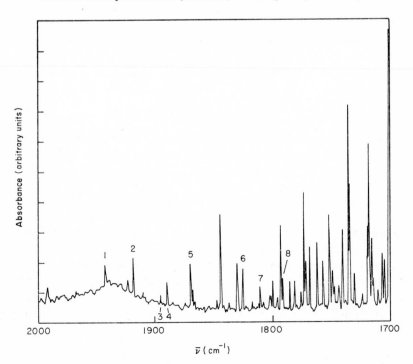

FIG. 4.6. Absorption of water vapour.

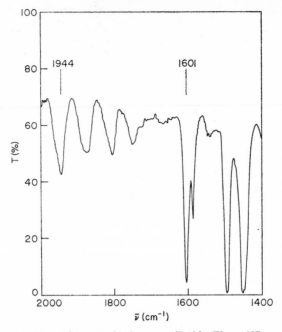

FIG. 4.7. 1944 and 1601 cm^{-1} bands of polystyrene (Perkin–Elmer 457 spectrophotometer, medium scan, normal expansion).

Some Effects of Instrumental Error

Error in the baseline is not too serious in its effects, provided the baseline is roughly flat. The relationship between true and observed sample transmittance is

$$T(\text{obs}) = T(\text{true}) \times T_0 \qquad (4.6)$$

where T_0 is the transmittance recorded in the absence of sample; similarly for absorbance

$$A(\text{obs}) = A(\text{true}) + A_0. \qquad (4.7)$$

Error in the transmittance zero is more serious, as Fig. 4.3 above and the accompanying discussion show. Such an error can arise from variations in the amount of stray light, of frequency not directly related to that demanded by the monochromator, that nevertheless reaches the detector. Unfortunately, it is not possible entirely to compensate for stray light by instrument adjustments, since some light may come through the sample itself, and thus be unavailable in the conditions under which the zero transmittance adjustment is carried out. Moreover, the amount of stray light entering through the sample depends on its absorption spectrum at frequencies other than that under investigation. Stray light is most troublesome when the true signal is weakest and will thus most affect strong bands. Fortunately, the region of greatest interest to carbonyl chemists is one where the genuine source output is high.

Insufficient pen sensitivity leads to sluggish response, and a tendency for the pen to fail to return to the baseline. Thus all fine detail is lost and the spectrum is almost completely valueless. This fault may sometimes be treated by increasing the gain, although usually the cause lies deeper, in the deterioration of some component, loss of alignment, or dirt on the mirrors.

Excessive pen sensitivity is caused by too high a gain together with high scanning speed. On returning to the baseline after recording a strong absorbance peak, the pen overshoots and may show a series of damped oscillations. As a result, information about the shape of the band, and especially about low-energy shoulders and ^{13}CO satellite peaks, may be lost.

Excessive scan speed leads to errors in the measured intensities, positions, and shapes of strong sharp bands. The maximum of absorption is traversed before the pen can fully respond, so that peaks appear weakened, displaced in the direction of scan, and with a shape that owes as much to the characteristics of the instrument as to the true form of the bands.

4.3. Sample Preparation and Handling

The light-absorbing properties of any molecule depend on its environment, so that the spectrum of a sample will depend on the phase in which it is recorded. The physical form and particle size of solids can also affect the observed properties. These considerations apply more forcibly to metal carbonyls than to most non-ionic materials, since the strong absorption bands correspond to large oscillating dipoles. In addition, many carbonyls show marked instability, especially in solution, some tendency to decompose in the sample beam, or to react with solvents, and moderate to high sensitivity to air oxidation. In this section we describe the various methods of sample preparation available, the precautions that must be taken, and the effects of some sources of error.

Cell Window Materials and Handling

In most cases samples are mounted inside some kind of cell, with windows transparent to light of the frequencies of interest. Cell window materials are generally ionic solids, which are transparent above a certain frequency, although for low frequency work polythene is useful. Some commonly used cell window materials are described in Table 4.2.

TABLE 4.2. Transmittance range of some window materials.

Substance	High energy limit (cm^{-1})	Low energy limit (cm^{-1})
Quartz	UV	2300 (a)
Sapphire	UV	1650 (b)
Calcium fluoride	UV	1100
Sodium Chloride	UV (c)	550
Potassium Bromide	UV (c)	300
Silver Chloride	UV (d)	450
KRS-5 (e)	20,000	250
Polyethylene (f)	500	10

(a) Useful range depends on grade of quartz.
(b) At liquid nitrogen temperatures. Room temperature performance slightly inferior.
(c) High energy limit will in practice generally be determined by scattering.
(d) Darkens when exposed to light of energy above 20,000 cm^{-1}.
(e) Thallous bromide–thallous iodide.
(f) Thin films or bags useable through much of conventional IR range.

The most commonly used window materials are sodium chloride and, for longer wavelength studies, potassium bromide. The windows, which are cut from single crystals, are brittle, easily scratched, and sensitive to moisture. They must therefore be handled with care, never allowed to rest except on a clean, soft, dry surface, and periodically polished with a suspension of rouge in methanol. Ideally, the operator should wear rubber fingerstalls, and in any case the optical surfaces should not be touched by hand.

Silver chloride windows are insensitive to moisture, but are gradually darkened by visible light, and exposure should be minimised. Windows made from thallium salts (poisonous!) do not suffer from this defect. Other water-insensitive materials include calcium fluoride and synthetic sapphire. These have high cutoff frequencies and are very expensive, but are useful for special applications (Section 4.4).

Vapour Samples

The form of the observed infrared bands of vapours is distorted by rotational structure. Despite this complication, it may be desirable to obtain vapour phase spectra, especially for compounds that are known to show isomerism, in order to minimise the interaction between a molecule and its environment.

The simplest vapour cell is a hollow tube, with windows at either end, and a sidearm stopcock adaptor that can be connected to a handling line. More sophisticated cells may carry a sidearm that can be loaded with sample.

After loading, the cell is evacuated, and disconnected from the vacuum line so that the sample material volatilises. There may be provision for heating the cell and the sample. If so, it is important that the sample material is heated less than the cell itself, to minimise condensation on the walls. Condensation on the windows is likely to be a problem whenever the sample is heated, although the infrared beam itself does have some warming effect. There is always the risk of decomposition on warming; this can be detected by the appearance of the spectrum of free carbon monoxide. Decomposition will not only weaken the genuine sample bands (except if the vapour is maintained at saturation pressure) and superpose the spectrum of CO and possibly of other decomposition products, but will lead to the formation of deposits of metal on the cell walls and windows. An alternative to raising the temperature of the sample is increasing the pathlength, and "multiple pass" cells (Fig. 4.8) are available giving pathlengths of up to several metres.

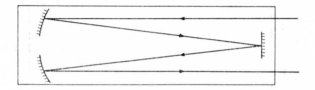

FIG. 4.8. Diagram of multiple pass cell.

Solid Samples

In this section we consider only the routine methods available for the study of polycrystalline solids. The special methods required for single crystal studies are considered further in Section 4.4 below.

Solid or glassy films may be deposited on plates by the evaporation of solutions. Unless the films are truly glassy throughout, they will consist of crystalline regions separated either by gaps or by amorphous regions.

The ability of a crystal to absorb infrared light is shown in Chapter 6 to depend on its orientation in the beam. Thus the spectra of aligned films differ from those of the component molecules. If the crystalline regions are of varying thickness or separated by gaps, then a completely misleading picture of the relative intensities of bands may be given. This is because the light recorded by the spectrometer is the sum of that passing through both sample-rich and sample-poor regions of the plate (Figs 4.9, 4.10).

Pellets or discs of sample diluted with a non-absorbing material are more widely used; suitable diluents are sodium chloride and potassium and

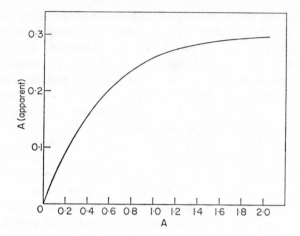

FIG. 4.9. Apparent absorbance of sample with 50% coverage.

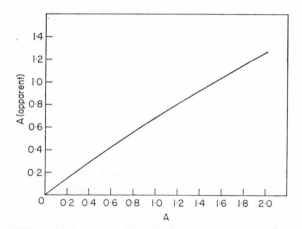

FIG. 4.10. Apparent absorbance of sample with 50% coverage to nominal sample thickness and 50% coverage at half nominal thickness.

thallium bromides, though polythene is useful in the far infrared. The sample material is finely ground, as is the diluent, and the two are intimately mixed. Fine grinding is essential if the sample is to be uniformly distributed throughout the disc, and may be accomplished mechanically (using a ball mill, or an automatic pestle and mortar) or with a pestle and mortar by hand. Coarse grinding will lead to loss of intensity through scattering, and of resolution through sample non-uniformity.

The diluted powdered sample is then placed in the barrel of a die-press, where it is subjected to pressures of between 1,000 and 4,000 atmospheres.

The material should be spread uniformly over the base of the press, or the pellet will be brittle and inhomogenous, and die wear will be increased. Before and during the actual pressing of the disc, the region containing the sample material is connected to a rotary pump, otherwise pockets of air will be trapped under pressure and the disc will shatter when the pressure is released.

The processes of grinding and pressing involve heating of the sample, in contact with the diluent. This may lead to thermolysis, thermal rearrangement, or chemical reaction. The need to evacuate during pressing leads to loss of volatile materials. Where polythene is used as a diluent, the sample particles sometimes become electrostatically charged, and concentrate in the outer part of the disc.

Mulls are probably more used than any other arrangement to obtain routine spectra. Unfortunately, these spectra are always to some extent ambiguous, as it is not clear to what state of the sample material they refer. A mull is prepared by mixing the sample material, as a finely ground solid, with a viscous liquid diluent, the function of which is to reduce losses by reflection and scattering, and to hold the sample material in place. Common mulling agents are nujol (heavy mineral oil), hexachlorobutadiene, and fluorocarbons. The mulling agent chosen will depend on the region of the spectrum to be studied, since mulling agents have spectra of their own (Fig. 4.11). Nujol may contain water or unsaturated hydrocarbon. Hexachlorobutadiene and some samples of fluorocarbon can react with some carbonyls hexachlorobutadiene suffers from the further disadvantages of fairly low viscosity and high volatility. Mull spectra of materials totally insoluble in the mulling agent are true spectra of polycrystalline solids, while spectra of very soluble materials are, in effect, solution spectra. If the sample is of intermediate solubility, mull spectra may contain parts due to solid and solution superposed, and the appearance of the spectrum may depend on the dilution of the mull and hence may vary from one specimen to another.

Mulls are generally applied to a plate with a spatula and held in place by a second plate. Holders are commercially available (Fig. 4.12), and tend to require the use of 25 mm diameter plates. However, 12·5 mm diameter plates may be used with very little loss of beam height, and a holder design used for these in the author's department is shown in Fig. 4.13.

Suitable mull dilutions and sample thicknesses are found by trial and error. If too much sample is present, the pathlength may be reduced by pressing the plates closer together, and the mull may be thinned *in situ*. Plates used for collecting the spectra should be rinsed with chloroform after use and left resting on beds of tissues or cotton wool in bottles full of fresh

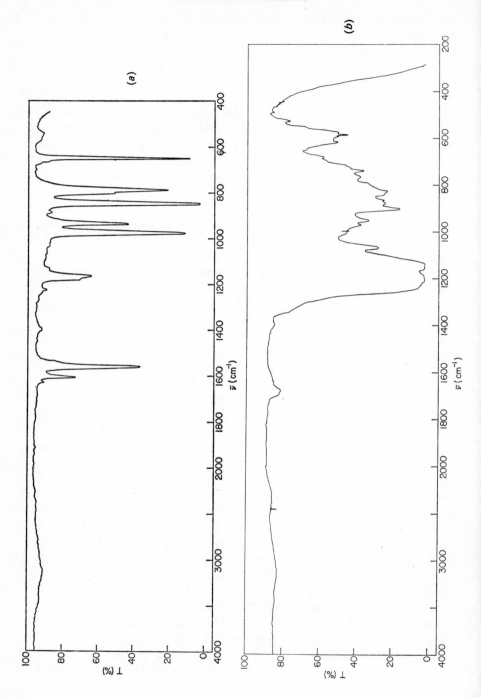

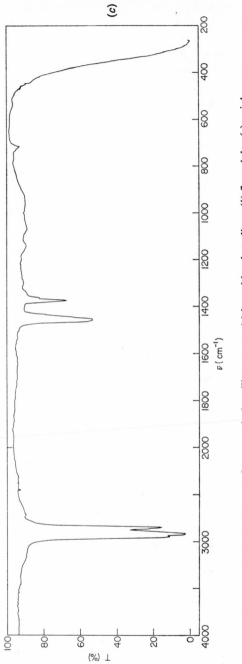

FIG. 4.11. Spectra of some typical mulling agents; (a) hexachlorobutadiene, (b) fluorolube, (c) nujol

desiccant. Repolishing will inevitably be necessary from time to time. If sample material is very scarce, it may sometimes by mulled by grinding between the plates, but the intervals between repolishing will be drastically reduced. It is almost impossible to balance a mull sample with a reference containing an identical amount of mulling agent, and it is best to collect mull spectra with nothing in the reference beam, adjusting the baseline to a convenient value in a region where neither sample nor mulling agent absorb, and distinguishing between sample and mull peaks with the help of reference spectra.

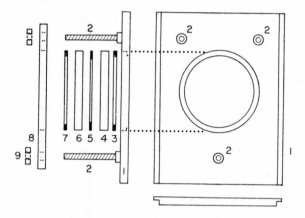

Fig. 4.12. Standard holder suitable for 25mm plates. (1) Rear holding plate (2) Fixing bolts (3,7) Gaskets (4,6) Plates (5) Spacer (8) Front holding plate (9) Fixing nuts.

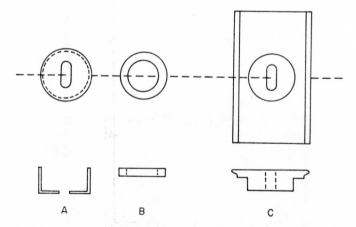

Fig. 4.13. Holder suitable for 12·5mm plates, which sit within ring B, held firm by cover plate *A* against the base of locating plate *C*.

Solutions and Liquids

It should be clear by now that the most informative spectrum of a substance is to be obtained from its spectrum in solution, provided that the interaction between solvent and solute is small. This proviso is particularly important for spectra in the carbonyl region, where the absorption spectrum of interest gives rise to a large oscillating dipole, involving atoms that are electronically and sterically unsaturated and located on the outside of the molecule.

Choice of solvent. The best solvents for spectroscopy are the worst from the point of view of solubility, since they interact most weakly with solutes.†
The interactions are to some extent specific. A basic solvent will interact most strongly with unsaturated metal complexes, while a solvent of high high dielectric constant will tend to dissolve, and interact with, highly polar compounds or salts, and a hydrogen bonding solvent may interact specifically with carbonyl anions. Both the average strength and the variability of the solvent–solute interaction are important, since such variability leads to peak broadening. A rough order of increasing solvent unsuitability in the absence of any more specific effects, is

saturated hydrocarbons < unsaturated hydrocarbons < ethers

< alcohols < carbon tetrachloride < methylene chloride

< chloroform.

The mixing of dissimilar solvents should be avoided, as it has been shown that this can lead to peak broadening or even splitting, due to the range of different solute environments that become possible.

An obvious point, but one sometimes overlooked, is that solvents may themselves react with the sample material, as may cell windows. Highly chlorinated solvents may oxidise the sample, and in particular may cleave metal–metal bonds [71]. Basic solvents may cause disproportionation [72] or ionisation [73]. Halide anions in cell window material may act as nucleophiles, particularly if polar solvents are used, and displace metal carbonyl anions that are good leaving groups [74].

Most commonly used solvents are fairly transparent in the carbonyl region, but solvent absorption can of course be a problem in other parts of the spectrum. The spectra of some typical solvents are shown in Fig. 4.14. It is obvious that solvents for spectroscopy should be pure, since impurities detract from transparency, and where such labile species as metal carbonyls are concerned there is the additional complication that the impurities may react

† The self-interaction of a solvent is also important. Thus methylcyclohexane and decalin are better solvents than hexane, and toluene is a better solvent than benzene, presumably because solute cavities can be formed more readily.

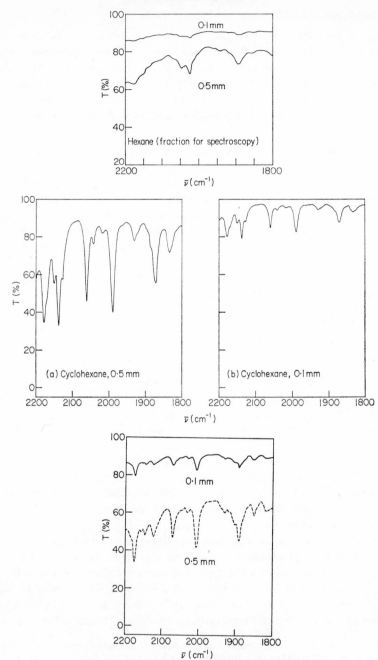

FIG. 4.14. Spectra of some typical solvents.

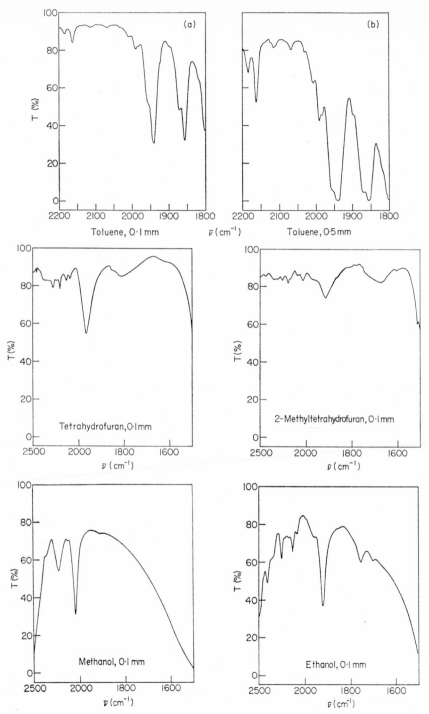

Fig. 4.14—*contd.*

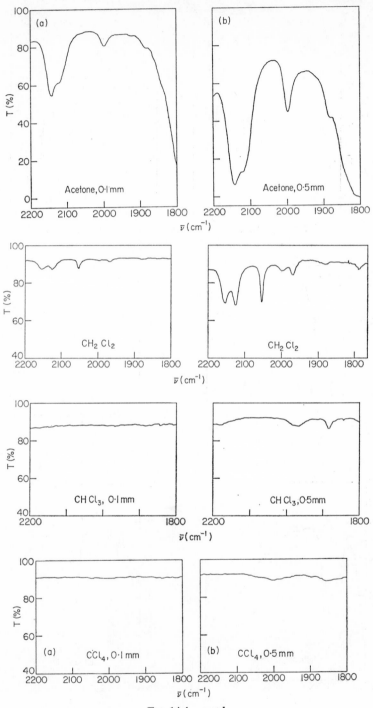

FIG. 14.4—*contd.*

with the solute. A universal impurity is water, which may generally be removed by zeolites. Ethers and some hydrocarbons are commonly contaminated with peroxides as a result of air oxidation, and these can be conveniently removed by distillation under nitrogen (care being taken not to allow dangerous concentrations of peroxide to build up in the distillation flask). Ethers in particular often prove troublesome, and should ideally be stored over sodium ketyl† and freshly distilled under nitrogen shortly before use. Chloroform contains a small amount of ethanol which is added to prevent phosgene from accumulating. This can readily be removed by passing the solvent through activated alumina or silica shortly before use.

Some metal carbonyl complexes, e.g. the *tri-n*-butylphosphinepenta-carbonyls of chromium, molybdenum and tungsten [75], are liquids at room temperature and may conveniently be examined as such. The substances may be thought of as dissolved in themselves. They are not particularly good solvents, from the spectroscopic point of view, although the long hydrocarbon sidechains help reduce molecular interactions.

Choice of pathlength and concentration. Up to a point, the shorter the pathlength the better, since solvent absorption is minimised. It is then necessary, however, for the solution to be more concentrated in order to get a given amount of solute into the light path. At small pathlengths, deviations of the cells from ideal geometry can become proportionately more important, and act as a source of sample inhomogeneity as well as of important relative error in the nominal pathlength. A convenient compromise for many purposes is a pathlength of 0.1 mm and a concentration around 10^{-2} molar in ligand CO.

The design, use and care of solution cells. A solution cell is a device in which two rectangular plates are held in position a fixed distance apart by a spacer, which also seals the gap between them. One of the plates has holes drilled in it, through which solution can be introduced by way of ports machined to accept a standard syringe coupling (Fig. 4.15). The spacer is generally either lead or teflon. A lead spacer has the advantage of greater dimensional stability, but can be made to give a reliable seal to most window materials only by careful amalgam treatment, and becomes useless if bent. Cells are classified as "demountable" or "permanent", depending on

† The ether, after rough drying, is placed in a stoppered *round-bottomed* flask and is treated with samples of sodium wire until these retain their brightness. Benzophenone is added, and the bright blue colour of sodium ketyl, $Ph_2CO^- Na^+$, appears. The flask is stoppered and shaken, and more sodium wire and benzophenone are added as necessary.

Sodium ketyl is an excellent reagent for the removal of oxygen, peroxide, and hydroxylic and acidic materials. If solutions of the reagent are stored in sealed vessels under air, a partial vacuum develops, due to depletion of atmospheric oxygen. The stopper should be sealed in place using a hydrocarbon grease, otherwise the alkali formed will cause sticking.

the ease with which they may be taken apart. Since metal carbonyls are often labile enough to deposit an insoluble film inside a cell, the windows will need cleaning and repolishing more often than cells used only for organic substances, and demountable cells are to be preferred.

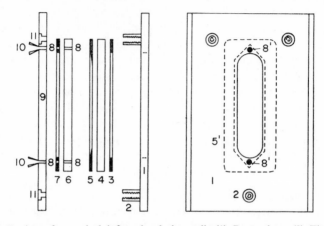

FIG. 4.15. Design of a typical infrared solution cell. (1) Rear plate. (2) Fixing point. (3) Rear gasket. (4) Rear window. (5) Spacer. (5′) Outline of spacer projected on (1). (6) Front (drilled) window. (7) Front (drilled) gasket. (8) Bolt holes (8′) Location projected on (1). (9) Front plate. (10) Inlet ports. (11) Channels for fixing bolts (bolts not shown).

To take a spectrum, the cell is filled from a syringe, its ports stoppered, and surplus solution wiped off. Care must be taken that no air bubbles are trapped between the plates at this stage. When the spectrum has been taken, the solution is forced out by clean solvent, which may in turn be forced out by a highly volatile solvent such as methylene chloride. The cell is then dried by sucking filtered laboratory air through it, or by passing through a dry gas from a cylinder. Compressed air should not be used, being contaminated with compressor oil. During this drying process, the cell should be warmed by an infrared lamp if, as is usually the case, the windows are made of a moisture-sensitive material. The cell is then returned to its desiccator.

Reference cells and baselines. Although it is unusual for a solvent to show strong absorption in the carbonyl region, weaker bands (due, for example, to combinations of HCH deformations) may be present. These can be made far less troublesome if a cell of the same material and pathlength as the sample cell, filled with pure solvent, is used as a reference. The cells may not be perfectly balanced even though their nominal pathlengths are equal, so that accurate work requires a baseline of solvent against solvent. Even if this baseline is straight, showing that the cells are in balance under these conditions, it should not be assumed that balance is perfect during the collection of the spectrum of a sample. The concentration of solvent is lower in the sample than in the reference cell, so that weak inverted features may appear where the solvent absorbs. Variable pathlength cells are available to improve balance, but their use is tedious, and it is simpler to use fixed cells, taking care to examine weak features with suspicion and with reference to the pure solvent spectrum.

The "nujol spacer" technique for routine spectra. Solution cell windows are more expensive, more fragile and more difficult to polish than the circular plates used for mull spectra, and the cleaning of solution cells can be a tedious business. Where a large number of spectra are required, but accuracy is unimportant, it is tempting to sample a solution simply by withdrawing a drop on the end of a glass rod, and to examine it as a film held by capillary action between plates. Under these circumstances the solution is liable to evaporate in the heat of the beam, or to become more mobile and run out from between the plates, and is also exposed to atmospheric attack. A more reliable but hardly less convenient method is the "nujol spacer" technique. In this, a very thin film of nujol is smeared over the plates before they are wetted with solution. The nujol then forms a spacer round the outside of the sample and protects it. It should of course be remembered that the observed spectrum includes peaks due to nujol and solvent, as well as to the carbonyls.

Monitoring reactions. The metal carbonyl region is extremely useful for the monitoring of reactions, whether these are novel or well-established. Not only does the appearance of new peaks show that a product is being formed, but their intensities and positions give useful information about its possible nature.

The simplest method for routine monitoring is the nujol spacer technique described above. The reaction may be one that must be carried out under nitrogen. If so, it should take place in a flask with one stoppered sampling port. When a sample is to be examined, nujol-smeared plates are prepared, and the pressure of nitrogen above the reaction is increased. The sampling port is then unstoppered and one plate held just above it. A drop of liquid is rapidly transferred on the end of a glass rod from the solution to the nujol-smeared plate, and a second plate is immediately placed over it. It is important to bring the covering plate to the sample-laden plate, and not vice-versa, so that the sample is never moved away from the protective draught of nitrogen that issues from the reaction vessel sampling port. With practice it is possible to remove fairly uniform samples in this way, so that the appearance and disappearance of peaks can be followed semi-quantitatively. Some very air-sensitive materials can be handled successfully, even by relatively unskilled operators; one successful undergraduate exercise in the author's Department involves monitoring of the reduction of dimanganese decacarbonyl to $Mn(CO)_5^-$.

When a reaction is being studied carefully, a more quantitative approach becomes necessary. A suitable procedure is to withdraw a sample through a side-port of the reaction vessel into a syringe, using a protective stream of nitrogen gas, and to collect its spectrum in a solution cell. A quantitative plot of absorbance against time should be kept for all peaks. This is to facilitate the assignment of different peaks to common species, since if they do belong to the same compound the ratio of their absorbances should be constant.

Errors Due to Defects in Sample

Inhomogeneous samples will lead to an under-estimate of the differences in intensity between different peaks (Fig. 4.9), and can thus lead to quite gross distortion of the general appearance of a spectrum.

Too much or too little sample. The effects of too little sample in the light beam are obvious. Peaks will be weak and all but the strongest may escape detection. If too much sample is present, even moderately weak peaks will appear strong and broad, so that it will not be possible to form any idea of band shapes, and strong peaks close together may not even be separately resolved.

Uncompensated solvent peaks. These may be present even in careful solution studies, and will certainly be present in rough reaction-monitoring work. Inverted solvent peaks may be present if a reference cell is used. In the carbonyl region, fortunately, most solvent peaks do little harm providing their possible presence is recognised.

Solvent cutout. If attempts are made to collect properly solvent-compensated spectra in a region where the solvent absorbs, two things will happen. Spurious normal or inverted peaks will be visible at the edges of the absorption band, due to imperfect cell balance, and within the band the pen will lack power, so that it may drift and generate totally meaningless features.

Sample decomposition. This problem can be reduced by avoiding highly chlorinated solvents, and by using freshly prepared solutions or mulls. The spectrometer beam itself supplies heat to the sample and this can increase the rate of decomposition, therefore it is prudent to rerun spectra of suspect samples after leaving them for an interval in the beam. It may also prove helpful to use a cell maintained below room temperature, while for very air-sensitive materials special techniques are necessary; the relevant procedures are described in Section 4.4 below,

4.4. Special Techniques

Air- and Moisture-sensitive Materials

All metal carbonyl compounds are metastable to air oxidation or to hydrolysis. Often the reactions are too slow to be detected, but where this is not the case, special care is necessary to prevent contact between the sample and air. This may be done using handling lines or dry boxes.

If a substance is sufficiently volatile, it may be transferred within a vacuum line using temperature differentials. Solutions may then be prepared by condensing solvent onto the sample material. The solution may be made up in a flask with a side-arm sealed by a rubber cap. To collect a spectrum, an inert gas (generally nitrogen or argon) is admitted to this flask to bring it to atmospheric pressure, and a generous sample is withdrawn through the cap by way of a demountable needle into the body of a syringe. Special long needles for Luer syringes are obtainable from medical suppliers, and are highly suitable for this purpose. The syringe needle is then removed and the solution forced into the solution cell, thereby flushing it of air and of initial oxidised material. Contact with air within the dead space of the syringe and the cell may be minimised by pre-loading with dry nitrogen or with solvent. It is advisable to err in the direction of excessive sample concentration when making up the solution, since it is then fairly easy to condense further amounts of solvent into the solution flask. The solution flask may with advantage be graduated and also contain a magnetic stirrer, this will facilitate quantitative dilution. Absolute sample concentration can be determined by withdrawing a known volume of solution, quantitatively decomposing the carbonyl complex (by shaking with bromine water, for example), and analysing for metal by standard methods.

The need to transfer the solution through the atmosphere does not arise in the dry box technique, which can also be applied to involatile materials. Dry boxes are of highly varying degrees of sophistication, but all consist basically of a sealed transparent-walled chamber. The contents of the box are accessible to an outside operator using rubber gloves built into the chamber wall. The chamber is filled with an inert gas (usually nitrogen or argon) at slightly above atmospheric pressure. This gas is continuously bled in or (in more sophisticated versions) recirculated over desiccants and deoxygenators. Objects are introduced into the dry box through a sample lock. Pieces of apparatus that can trap air readily, such as cells and syringes, should be flushed with dry box air several times when first introduced to the box and again just before use. The dry box technique is suitable for solids or liquids, and solutions may be made up quantitatively inside the box, although weighing inside the box can be made more difficult by static charges. Mulls may be prepared and placed between plates, and protected from subsequent air oxidation by sealing the outside rim of the plate assembly with grease. Solution cells may be filled and sealed. Vessels may be attached to handling lines through stopcock adaptors, used to collect samples or solutions, and then isolated and transferred to a dry box for subsequent processing. Thus the dry box technique posseses the advantage of flexibility; but the purging and re-purging required can be time-consuming.

Temperature Variation Studies

Studies over a range of temperatures give information about reversible processes, particularly isomerisation and rotamerism, and about "hot bands" (such as difference bands) due to the absorption of light by vibrationally excited molecules. In addition, low temperature spectra are usually sharper, and thus more informative, than comparable spectra obtained at room temperature. There is some evidence [76] that absorption maxima at room temperature are less directly representative of the molecule than are low-temperature values, since the bands are in fact envelopes for a

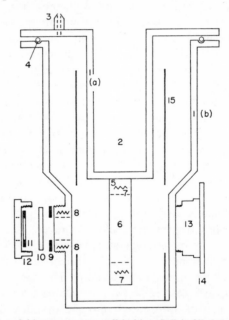

Fig. 4.16. A typical variable temperature cell holder. Cell is filled through threaded ports sealable with microbolts. (1) Cell body; (a) inner, (b) outer. (2) Refrigerant well. (3) Lead to vacuum. (4) O-ring sealing 1(a) to 1(b) (bolts clamping 1(a) and 1(b) omitted for clarity). (5) Cell block. (6) Cell space. (7) cell block heaters. (8) Outer port window heaters. (9) Gasket. (10) Outer window. (11) Gasket. (12) Window assembly ring. (13) Window assembly (parts 8–12, not individually shown). (14) Cell holding plate. (15) Radiation shield. A commercially available cell of this pattern, (Beckman VLT-2) has self-sealing ports for leads in the upper part of 1(a).

whole range of species in which various low frequency vibrations have been thermally excited.

A typical variable temperature cell, useful from − 185 to + 250°C, is shown in Fig. 4.16†. The sample is in thermal contact with a copper block, the temperature of which is controlled by the contents of a reservoir. If so desired, the temperature of the block may be further adjusted by the use of ancillary heaters. Thermal contact between the sample and the block may be improved by generous amounts of silicone grease, but it is important not to let this contaminate the specimen. Liquid specimens can be examined in a demountable cell, with ports designed to be sealed tight against vacuum. This inner assembly is insulated by a vacuum chamber with transparent windows. Heaters minimise condensation of moisture on these outer windows in low-temperature studies. The outer windows may be made of any convenient material. Sodium chloride is suitable for the windows of the inner solution cell in high-temperature studies, but tends to become clouded with moisture or to crack in low-temperature work, for which silver chloride is recommended. The gaskets and washers of the variable temperature cell may be of lead or teflon. In either case, temperature changes inevitably lead to changes in the relative dimensions of the cell components. The technique of assembling a cell so that it remains leakproof over a range of temperatures is one that requires practice.

For high temperature studies, the solvent chosen must be one that does not develop an excessive vapour pressure. For low temperature work the solvent must be one that stays liquid down to the lowest temperature of interest, or that forms a glass. The solvent must also be carefully dried to prevent light scatter by ice crystals; phosphorus pentoxide should not be used, as it gives rise to much the same problem as does dissolved water.

Temperatures down to 20 K may be obtained using the Joule–Thompson cooling of hydrogen to provide refrigeration to the sample holding block, in apparatus of the type shown in Fig. 4.17.‡ There is no intrinsic difficulty about building cells to operate at temperatures down to 4·2 K, although somewhat more complex apparatus is required, incorporating a liquid helium refrigerant reservoir. This reservoir and the sample are both protected from as much thermal radiation as possible by a shield cooled with liquid nitrogen. These methods may be applied to solid samples or to the glassy solutions and condensed vapour matrices discussed below.

A problem in all low temperature solution studies is poor solubility. Since precipitation is itself a physico-chemical process, it may be inhibited in

---

† Cells of this basic design are commercially available from Beckman-RIIC Instruments.

‡ Commercially available microrefrigerators of this kind include the 'Cryotip' series, obtainable from Air Products Inc., and the Cryodyne series, produced by Cryogenic Technology, Inc.,

some cases by cooling as rapidly as possible. High solvent viscosity will also slow down precipitation, and the ideal low-temperature solvent is one that can be made to form a glass by cooling before precipitation takes place.

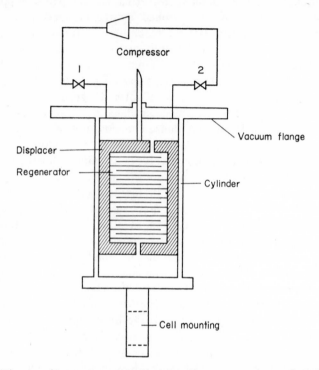

FIG. 4.17. Diagram of low-temperature cell operated by gas expansion cooling. Compressed gas is admitted through valve 1 into the barrel of the cylinder, with the displacer (a long hollow thermally insulating cylinder) at the bottom of its stroke. The displacer is raised, causing the gas to flow over the heat-exchanging regenerator within and be cooled. Valve 2 is opened, allowing the gas to expand and cool the assembly. Finally, the displacer is returned to the bottom of the cylinder, thus re-cooling the regenerator material.

Glassy Solutions

The main advantages of glassy solutions are described above. They enable the effective range of solution studies to be extended down to temperatures where the solvents are rigid although equilibrium solubilities may be minimal. In addition, they are extremely useful for the study of unstable species, especially those generated photochemically.

Most pure solvents freeze to crystalline solids on cooling, and in the process deposit crystals of the sample material. Some solvents or solvent mixtures behave differently, becoming rigid (glassy) without crystallising.

The solute is then frequently trapped in the glass as a metastable solution, ideal in some ways for infrared studies. Some suitable solvents for glass formation are isopentane–methylcyclohexane mixtures, mixtures of ethanol and methanol, and 2-methyltetrahydrofuran. For the first of these at least, nujol may be added to the mixture. It has the effect of increasing viscosity, raising the glass point, and thereby greatly reducing the degree of precipitation [77].

The spectrum of the solvent itself changes at the glass point, the weak combination bands that overlap the carbonyl region becoming sharper and of higher extinction coefficient. This renders a baseline essential, and a reference cell containing glassy solvent is highly desirable. In some cases, the solute spectrum broadens below the glass point, presumably because of trapping of the solute in a large number of more or less distorted environments. The form of the bands may sometimes be improved by annealing. Bands may be split by at least two mechanisms. The glass may contain more than one kind of site for solutes, and these may lead to different spectra, or there may be only one kind of site, in which solute molecules are subjected to a solvent field of low symmetry. Impurities may be trapped in the same cage as solute molecules, leading to an environment that is not only assymmetric but untypical of the solvent.

Despite these disadvantages, the glassy solution method is relatively easy to use and requires no very expensive apparatus (at least down to liquid nitrogen temperatures). It may be applied to any materials, including ionic materials, that are sufficiently soluble in a suitable glass-forming solvent. The trapping of different materials in the same solvent cage can be turned to advantage in studies of molecular interactions.

Matrix Deposition from Vapour

The method of matrix of deposition from vapour may be applied to a wide range of materials with sifficient vapor pressure, and since $Fe_2(CO)z$ has been successfully studied in this way [78], the minimum vapour pressure required must be quite low. The sample vapour is diluted with a suitable inert gas, and the mixture is sprayed through a nozzle onto a transparent cooled support, such as potassium bromide window. For reasons that are not obvious, intermittent pulsed spray-on gives results superior to those obtained with a slower, steady stream of vapour. Sample to diluent ratios of around 1:1,000 are commonly used.

The noble gases, nitrogen, methane, sulphur hexafluoride and carbon monoxide have all been used as carrier gases. Noble gas matrices enable the entire spectral range to be studied. Spectra may readily be obtained at all temperatures from 20 K (with a hydrogen expansion refrigerator) up to the temperature at which the carrier gas in the cell begins to sublime. The condensation process must involve a very rapid loss of energy from the condensate, and presumably the formation of a highly defect matrix. This can lead to local asymmetries of environment for the material under study, and to small distortions and splittings in degenerate bands. To some extent, these effects can be modified by annealing [44]. Rapid condensation has, however, the advantage that the test substance does not have the chance to aggregate, and that its molecular form is likely to be that in the vapour, unperturbed by solvent effects.

Unstable Species by Vapour Deposition

Species that cannot normally be isolated have been characterised by a version of the vapor deposition technique. For example, a range of carbonyl fragments containing fewer than the normal number of CO groups can be prepared by co-condensing elemental metal vapour, carbon monoxide and an inert carrier gas. This method has even been used to prepare carbonyl complexes of non-transition metals [79]. Vibrational spectroscopy is uniquely useful in assigning possible structures to the fragments formed, which would not otherwise be characterisable. The products obtained can be varied and assignments thereby tested by altering the ratio of carbon monoxide to carrier.

Photochemically Generated Species

Photochemical substitution according to a general scheme

$$M(CO)_n \xrightarrow{h\nu} M(CO)_{n-1} + CO$$

$$M(CO)_{n-1} + L \rightarrow M(CO)_{n-1}L$$

is a process much used in synthetic metal carbonyl chemistry. The final product, which need not be isolated from solution, can if sufficiently stable be transferred to a cell using the general techniques described for air-sensitive materials.

More specialised techniques are necessary for the study of the intermediate photolytic fragments, or of materials that are only stable for appreciable lengths of time below room temperature. These techniques involve the photolysis of material that is already inside the infrared cell.

Photolytic fragments may be studied either in glasses formed by cooling solutions, or in vapour deposited matrices. In the former case, the choice of a window material for the low temperature cell requires care, since the usual material, silver chloride, is light-sensitive. The author has found that sodium chloride and calcium fluoride tend to crack under the strain that develops below the solvent glass point, and has therefore used synthetic sapphire. This is not cheap, and suffers from the major disadvantage of absorbing strongly below $1650 \, cm^{-1}$ at $-185°C$ ($1700 \, cm^{-1}$ at room temperature). Further limitations on solvent glasses in the study of photolytic fragments arise from the highly unsaturated nature of the species of interest, which react with donor solvents even at liquid nitrogen temperatures [45, 80], interact strongly with hydrocarbons [81], and are coordinated by molecular nitrogen [82].

Low temperature solution cells are, however, highly suitable for the study of moderately unstable photochemically generated species in solution. Such species may be formed by irradiating the parent complex in the presence of a ligand at temperatures slightly above the glass point, thus avoiding the problem of window fragility. Alternatively, the photolytic intermediate may be generated in a matrix that contains L. The desired product may then be formed either directly (if the metal carbonyl and L are in the same solvent cage) or on softening of the glass. In either case, care must be taken to distinguish the bands due to the desired product from other bands, such as those due to photochemical processes not involving L.

The matrix deposition technique is often more suitable for the study of photolytic fragments, provided the parent compounds are sufficiently volatile. There are no problems with window materials as such, and irradiation can be through a separate port from those used for collecting the spectrum. The noble gases, much used in matrix deposition studies, may be presumed to interact less strongly with the fragments than would any other materials. However, the process of matrix deposition precludes the preferential trapping of carbonyl and potential ligand in a common cage, and the matrices cannot be studied above their melting points. For these reasons, matrix deposition is less suitable than glassy solvent formation for studies of the fates of photolytic fragments.

Two other points should be mentioned that apply to both techniques. Firstly the range of photolytic fragments observable is limited by matrix cage effects. For example, the failure to observe loss of phosphines from metal carbonyl phosphine complexes, is due to these, since such loss can be demonstrated in solution at room temperature. A bulky ligand is unlikely to carry away enough surplus kinetic energy to allow it to melt its way out of the cage, and so the initial loss of such a ligand will be masked very effectively by recombination. Secondly, the products of photolysis may themselves be sensitive to light. This leads to the phenomenon of secondary photolysis, in which the initial products are photolysed further, and much experimentation with sources, filters and sample concentrations may be necessary to control this effect. The primary photolysis process can also undergo light-induced photoreversal, in which absorption and thermal degradation of light by the fragment causes enough local disturbance for recombination to occur.

Under some circumstances, a steady state is set up when photolysis is far from complete. This problem can be reduced by lowering the sample concentration, and by increasing the frequency of the exciting light.

Quantitative Analysis and Kinetic Studies

In principle, infrared optical densities can be used directly as a measure of concentration in quantitative or kinetic studies; indeed, the strong, well-resolved carbonyl bands are ideal for these purposes. In practice, errors can arise due to cell imperfection and finite optical slit widths. Correct setting of the zero of transmittance is, of course, critical in quantitative studies, and purging to remove water vapour is desirable.

Actual cell pathlengths must always be suspected of differing from their nominal values, and true values may be found interferometrically.† Moreover, the cell may not be of uniform thickness, so that the nominal absorbance will be an average of values for different pathlengths, and relative absorbance values will be distorted as in Fig. 4.9. The effects of optical slitwidth (Fig. 4.2) will be similar to those of cell non-uniformity, since of the light collected nominally at a band maximum, some will be of less strongly absorbed frequencies. Reduction in slit width leads to increase in noise as a proportion of signal, and hence to a degree of subjectivity in the measurement of absorbance values. A case could be made for deliberate band broadening by choice of solvent, but this could lead to overlap problems, and in any case the solvent is often specified in advance by the nature of the investigation.

These difficulties can be accommodated by an empirical "Beer's Law plot", not necessarily linear, of measured absorbance against concentration, obtained using a pure sample of the material of interest. A pair of cells should be assigned to the investigation, and care should be taken to standardise the spectrometer settings. The apparent absorbance at zero concentration need not be zero, since the cells are not necessarily perfectly balanced, so the zero reading should be given the same weight as all others. The range of concentrations used for the Beer's Law plot should cover all those that it is planned to measure.

It is not possible to measure accurately an absorbance that is either too high or too low. If the absorbance is low, then noise, baseline irreproduci-

---

† A spectrum of the cell filled with air is recorded. The light from this cell is a mixture of direct and multiply reflected beams, so that destructive interference occurs at some wavenumbers. The spacing between apparent absorption maxima is given by

$$\bar{\nu} = 1/2D$$

Multiple reflection depends on a high difference in refractive index between the cell windows and contents. For this reason, interference bands are not a nuisance in solution spectroscopy.

bility, and the uncertainty of reading, will lead to error, while high absorbances are seriously affected by finite slit width, and by the errors inherent in measuring the small amount of light transmitted. For these reasons it is advisable to confine quantitative studies within the absorbance range 0·1–1·0, and to use a range of cells of different pathlength if necessary, each with its own calibration plot. If a substance shows more than one carbonyl peak, both should be used where possible. This not only provides a check on all measurements, but may make it possible to avoid problems of overlapping absorption when other carbonyls are present.

For the analysis of mixtures, additional care is necessary. Infrared bands often have wide "wings" of measurable intensity, so a Beer's Law plot should be made for each component at the frequencies used to measure the concentrations of all the others. Calculating individual concentrations then involves the solving of a set of simultaneous equations, which, if the Beer's law plots are linear and corrected for any zero error, are of the form of Eq. (4.5). The assumption of additivity is not strictly justified if the non-linearity due to finite slit width is appreciable, but this error is minimised if for each component, bands can be found that are well separated from those due to the others.

Finally, some trivial sources of error should be mentioned. Errors can arise due to solvent evaporation, either during the preparation of standard solutions or while the sample is actually in the cell, especially if this is not rigidly airtight. The risk of reaction with solvent and with dissolved impurities is particularly great in the preparation of the solutions for the Beer's Law plot, since this involves two or more separate dilution steps. Thorough cell cleaning is of course essential, as is the regular checking of baselines, and, should it become necessary to replace or dismantle a cell, the entire calibration process must be repeated.

Intensity Studies

The Beer's law plots of the previous section are in effect measurements of the apparent molar extinction coefficient, and if it is possible to choose conditions so that optical slit width is far smaller than the width of the bands studied, the values obtained will be valid. Unfortunately, the extinction coefficient is not acceptable as a measurement of the intensity of a band. The width is also a factor, and depends on the form of the vibration that gives rise to it, on the strength of solvent–solute interactions, and on the degree of variability and disorder in the environments provided by the solvent. The only feature of fundamental significance is the *integrated intensity E*, defined by

$$E = \int_{\bar{\nu}=0}^{\infty} \varepsilon \, d\bar{\nu}. \tag{4.6}$$

If the effects of finite slit width can be ignored, then the integrated intensity of a well-resolved band can be determined by one of several methods. These include numerical integration, weighing an area of chart paper (which is a disguised form of numerical integration), and fitting the band to some assumed analytical form that can be integrated. Numerical integrations are necessarily confined to finite limits. Their results must therefore be corrected for the total "wing intensity" beyond these limits, which is estimated analytically assuming a particular form for the band. Whichever method of integration is chosen, a correction may be applied for the optical slit width using published tables [83], assuming a triangular slit function† and a Lorenzian† form for the absorption band.

Polarisation Studies

In disordered phases (gases, normal liquids and solutions), it is impossible to determine the orientation of the electric vector of light absorbed relative to the molecular axes. However, in ordered phases (solids and nematic solutions) it is possible to orient the sample in the laboratory, and extra information (see Chapter 6) may then be obtained by orienting the electric vector of the light, i.e. by polarising it.

Light may be polarised by selective reflection or by selective absorption. Selective reflection takes place most efficiently when the angle of incidence with which light strikes a plate of material is the Brewster angle, defined by

$$\tan \theta = \text{refractive index.} \qquad (4.7)$$

The oscillating electric dipoles associated with normal (unpolarised) light are oriented at random in all directions perpendicular to the direction of propagation. If a horizontal light beam strikes an upright plate of transparent refracting material at the Brewster angle, the light is resolved into two components. The component with the oscillating electric vector vertical is transmitted, while that with the vector horizontal is preferentially reflected.

The choice of plate material presents a problem, since refractive indices are generally low in the infrared region, and silver chloride and silver bromide are often used. These are photosensitive, and must be encased or else housed in the enclosed part of the spectrometer. A single plate does not discriminate sufficiently between polarisations, and a stack of plates must be used. This occupies a considerable pathlength, especially since it must make a very acute angle with the beam. Moreover, the stack must be located at a point in the beam separated from the sample only by slits and plane reflectors, since focussing the beam will alter its polarisation properties.

† These terms are defined in the discussion of the shapes of bands in Section 5.2 below.

For these reasons a polarising stack is compatible with some instruments only. If the cell compartment is long enough, an encased rotating stack may be used. If the stack is housed inside the body of the instrument, it may be slid in or out of the beam. With this arrangement it is easy to obtain unpolarised or vertically polarised light, but horizontally polarised light is not so easily obtainable. This is a serious limitation, as it may be difficult to rotate the sample, and doing so may involve losing part of the height of the infrared beam or changing the actual part of the specimen examined. Sacrificing any part of the beam intensity is a serious matter, since the transmittance of light of the favoured orientation (itself only half of that normally available) is far from perfect.

Selective absorption by wire grid polariser is both more efficient and more convenient in the infrared region. A wire grid polariser consists of a very fine array of parallel metal strips, produced by vapour deposition in vacuum onto a surface with machined parallel rulings of distance less than the wavelength of the light to be studied. A grid polariser is relatively small, is highly selective between polarisations, is not photosensitive and may therefore be placed in an accessible region, whatever instrument is used, and may be oriented at will. Its sole disadvantages are fragility, and expense.

Errors can arise in all polarisation studies due to the transmission of some light of the less favoured orientation, though with grid polarisers this should not be a problem. Frequency dependent polarisation preference in the instrument itself can cause spurious peaks, but this effect can be corrected for by using suitable baselines. A more serious source of error arises if the beam is not parallel, since the transmitted light will then include an unwanted component with an electric vector parallel to the nominal direction of propagation.

Nematic Solutions

A nematic is a liquid crystal with its molecules ordered in orientation, though not necessarily in position. Nematic phases are formed by a variety of materials, many of them azo-compounds, with rigid, highly anisotropic molecular structures. The nematic phase is thermodynamically stable only over a small temperature range, but may often be supercooled. If a highly anisotropic molecule is dissolved in a liquid which, on cooling, forms a nematic phase, the solute molecule will in many cases tend to adopt a particular orientation in that phase.

The use of nematic solvents then involves three steps. One of these, the polarisation of the infrared beam, is discussed above. The second step is defining the orientation of the nematic in the laboratory. A variety of very elaborate methods have been used for this [84], but recently it has been shown [15] that the orientation of one material at least (butyl *p*-(*p*-ethoxy-phenoxy-carbonyl)-phenyl carbonate) can be predetermined by stroking the plates between which it is allowed to cool, and that this material will persist as a supercooled nematic in the laboratory. The success of orientation may be verified from the polarisation dependence of the ultraviolet cutoff point of the solvent itself.

The third, and in some ways the most difficult step, is the choice of solute molecule. This should be long and rigid, so as to lie more readily in one direction than in another. More globular molecules are less likely to show a satisfactory degree of orientation, as are molecules containing long non-rigid side-chains. Thus $Mn_2(CO)_{10}$ shows a useful degree of order in nematic solution [15], but under conditions where this finding was reproduced no orientation of *tri-n*-butylphosphine chromium carbonyls could be detected [85].

Single Crystal Studies

In principle, a great deal of additional information about the absorption of light by solids can be obtained by studying single crystals using a polarised beam. In practice, this method is difficult to apply to metal carbonyls, because of the high extinction coefficients of the bands of interest; in order to give an acceptably low optical density, very thin crystals would be required. There does not seem much hope of obtaining spectra of carbonyls in host lattices, since the only possible hosts are likely to be other metal carbonyls. Detailed analyses of the spectra of solid carbonyls [86] have therefore depended on the spectra of potassium bromide discs, in conjunction with Raman studies, There may however be some hope of forming suitable crystals by evaporation of films of solution on transparent supports. Crystalline regions may then be detectable, and their orientation determined, using the polarising microscope [87]. If a suitable crystalline region can be found, then the remaining area of the support can be blacked out, and as much light as possible directed through the chosen crystal using a beam condenser (i.e. lens) system.

Isotopic Substitution

While ^{13}CO is present in carbon monoxide with 1.1% natural abundance, frequently this does not suffice for the observation of all the weak bands of interest due to singly labelled polycarbonyls, and the intensity of bands due to doubly labelled species is negligible. However, both ^{13}CO and $C^{18}O$ are available from a number of sources, and using these the amount of information available may be greatly increased.

The relative merits of partial and total substitution are discussed in Chapter 3; it is concluded that only total substitution gives results suitable for a full force-field calculation, and that only partial substitution gives information of value in determining energy-factored parameters†. The isotopically labelled gas may be handled by conventional methods on a line, and returned to its reservoir using a Toepler pump. The high volatility of carbon monoxide makes it difficult to return to a reservoir by cooling, but the use of an absorbant material is of some help. Isotopically labelled species may be prepared by carbonyl exchange, by reaction of labelled CO with a precursor, or by the chemical reactions of a previously labelled carbonyl. Carbonyl exchange may be either thermal or photochemical. Which ever method is preferred, it is more efficient to equilibrate the substance (usually in solution) with successive small portions of gas. It is helpful to collect spectra after each enrichment stage, to help in the assignment of different bands to species with different degrees of labelling. Equilibration of a substance in solution is very economical. No more material need be used than is required to collect a spectrum, since the same sample can be re-examined after successive enrichment steps. It must, however, be remembered that chemically different carbonyl groups may exchange (either thermally or photochemically) at very different rates. The reaction of labelled CO with a precursor can be more expensive in enriched CO, unless the reaction chosen is of very high yield. Chemical reaction of labelled material is likely to be the least efficient method of all, except for such highly accessible species as singly substituted carbonyl derivatives, which may be prepared photochemically in solution in high yield, and used without further isolation (See also Note, p.139).

High Pressure Techniques

Some metal carbonyl complexes show a range of equilibria with carbon monoxide and, in certain cases, with other gases such as hydrogen, and these processes presumably affect their behaviour when used as catalysts in industrial processes. Infrared spectroscopy is well suited to the investigation of the species formed, and special apparatus, in which carbonyls are examined in solution in the presence of gases at high pressure has been described, [88].

Low Frequency Methods

The CMC bending modes of metal carbonyls lie below $100 \, \mathrm{cm}^{-1}$, which places them far beyond the range of conventional infrared spectrophotometers. Source energies in this region are too low to give monochromated beams of acceptable intensity, and spectra are obtained interferometrically. In an interferometer, light from the source (typically a mercury discharge lamp, which gives more energy at low frequency than does a black body) is split into two portions, the path difference between which is varied by a moving mirror. The portions are recombined and passed through the sample to a Golay detector. For a given path difference the two portions of the

†Partial substitution fingerprints may also be used to decide geometry [349].

beam interfere constructively or destructively to different extents at different frequencies. The total energy reaching the detector is a function of the path difference, and from this function it is possible to compute the variation of detected energy with frequency. Comparison of results with and without a sample in position gives an absorption spectrum. Water vapour absorbs strongly in the region of interest, and the instrument is evacuated in use. Polythene is used as a cell window and lens material, and mirrors are front aluminised. The interferogram (as the plot of energy against path difference is called) must be converted to a spectrum by a Fourier transform; this operation is performed by a computer on the digitalised instrument output.

Combination Spectra

The binary combination and overtones of CO groups occur in the region $4,100–3,500 \text{ cm}^{-1}$, and ternary bands at correspondingly higher frequencies. The intensities of binary bands are far lower than those of fundamentals, while those of ternaries are lower still. Thus, long pathlengths (from 1 mm to 5 cm or longer) are necessary. The choice of solvent requires some care, since both saturated and unsaturated hydrocarbons generally show combination bands of their own in the regions of interest, and carbon tetrachloride and carbon disulphide have been used. The restriction on the range of solvents is unfortunate, since the interpretation of combination bands involves comparing them with fundamentals obtained *in the same solvent*. Careful drying of solvent is important, as is purging of the spectrometer with dry nitrogen, since water absorbs in the regions of interest. Stoppered silica cells, of the type generally used in visible and ultraviolet spectroscopy, should be used, and baselines collected. It is convenient to use a spectrometer with a 90–100% transmission scale to record weak peaks. Ternary combination bands lie outside the range of normal infrared instruments, but may be observed using the near infrared range of spectrometers designed primarily for higher energies, as may the binary combination bands. Calibration should be carried out for each individual spectrum, the calibrant and sample spectra being collected where possible on a single pass without interrupting the frequency scan. Suitable calibrants for the binary region are indene [70] and 1, 2, 4, trichlorobenzene [89] (Fig. 4.18). The ternary region may be calibrated using mercury and noble gas discharge lamps [70]. The temperature of the monochromator must be kept constant during each run, and spectra repeated to confirm the absence of spurious drift. If these precautions are taken, frequencies reliable to within 1 cm^{-1} may be obtained.

Combinations involving at least one mode other than a carbonyl stretch occur throughout the spectrum from $2,600 \text{ cm}^{-1}$ downwards. They may be of one of two different kinds. Sum bands arise when two vibrational modes are excited by the same photon; they may occur from the ground state of the molecule or from any other. Difference bands involve excitation of one mode while another is de-excited; they can only occur when light is absorbed by an excited molecule, and are therefore generally weak even in comparison with combination sum bands. Difference bands lose intensity on cooling, but the intensity of sum bands is almost independent of temperature. Despite their weakness, combination bands can be important as a way of locating inactive modes, both carbonyl stretching motions, and others, and are the only source of information about anharmonicity; these applications are discussed further in Section 6.1 below.

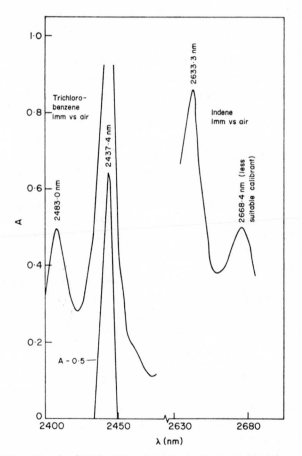

Fig. 4.18. Calibrant bands of 1,2,4-trichlorobenzene and indene in the binary $v(CO)$ region.

4.5. Some Remarks on Raman Spectroscopy

The technique of Raman spectroscopy has recently been authoritatively reviewed [22], and Raman instruments are still too expensive to be routinely available to most chemists. For these reasons, the present section is confined to a brief survey of considerations of special interest to carbonyl chemists. In modern instruments, the exciting beam is generated by a laser, which can be operated in combination with filters to produce one of a range of frequencies. Preference should obviously be given to a frequency that is not absorbed by the sample. Absorption leads to weakening of the exciting beam, and almost certainly of the Raman-scattered beam as well. Even more seriously, the absorption is commonly accompanied by photolysis, and by decomposition through sample heating. It may be possible to reduce sample heating by rotation or stirring, but the degree of photolysis will not be affected, although sample cooling may help with both problems. If photolysis is suspected, spectra should initially be collected using low exciting beam power. These spectra, which will have very poor signal-to-noise ratios, may then be used to validate spectra obtained under more normal operating conditions.

The problems of photolysis and absorption make it necessary to favour low exciting frequencies. On the other hand, the intensity of Raman scattering increases as the fourth power of the frequency. Moreover, photomultiplier sensitivity typically falls off below 20,000 cm^{-1}. This effect is all the more serious because of the frequency lowering of the Raman scattered light associated with the carbonyl modes.

Vibrations that are weakly Raman-active can sometimes be observed by the deliberate use of an exciting line in or close to a region of absorption (resonance Raman effect). This method has been used to find the skeletal breathing mode of $(C_5H_5)_4 Fe_4 (CO)_4$ [90]. One purely practical advantage of Raman spectroscopy deserves mention. The Raman experiment does not involve transferring material into a cell of critical dimensions, with all the handling problems this step entails for labile materials. Satisfactory spectra can be obtained using sealed tubes which can be brought into good contact with the optics of the spectrometer using glycerol or nujol.

4.6. Toxicity and Related Hazards

All metal carbonyls should be regarded as toxic, and in the case of nickel carbonyls, at least, the toxicity is far greater than that of the carbon monoxide they contain. The more volatile carbonyls are especially dangerous through inhalation, and should only be handled inside a good fume hood. Nickel carbonyl itself should be handled only inside a sealed line which is itself contained in a fume hood.

Fine crystals of cobalt carbonyl are pyrophoric, and other carbonyls (including $Fe_2(CO)_{12}$ and $Fe(CO)_5$) can deposit pyrophoric particles of metal.

Among solvents, carbon disulphide is unpleasant, toxic, volatile and highly inflammable. Carbon tetrachloride is poisonous through inhalation and skin absorption, as is benzene, and care should be taken to avoid spillage and in the disposal of residues. Wherever possible, toluene should be used instead of benzene. Not only is it less toxic but it is less volatile, has a longer liquid range at both ends, and is a better solvent.

Note added in proof:

The ultimate product of exchange between a metal carbonyl derivative and isotropically mixed CO is a statistical mixture of products. Such a mixture is generated directly in the matrix reaction of metal atoms with carbon monoxide. It has been pointed out that the spectroscopic properties of the general structure of the species under study and the ultimate intensity patterns for a range of species $M(CO)_n$ have been given (349).

It must be remembered that partial labelling always gives rise to a mixture of products, since molecules that have already acquired a label can nonetheless take part in further reaction. This feature can be turned to advantage, since the frequencies and intensity distributions of multiply labelled species provide fresh tests of any prepared force field.

Chapter 5

Interpreting a Spectrum

It is surprising how long it can take to learn to get all the information out of a spectrum. It is a skill that often seems to come only with practice, although the techniques involved are actually very simple. This Chapter contains some suggestions in the hope that the task of reading a spectrum will be made a little easier for the beginner. A few examples are included to illustrate the difficulties that may be encountered.

5.1. General Considerations

Measurement of Peak Position

Accurate measurement of peak position becomes far easier if an expanded wavenumber scale is used. The position of the absorption maximum can be measured directly, by reading off the wavenumber at which absorbance is greatest. However, this is not the most accurate method, since the rate of

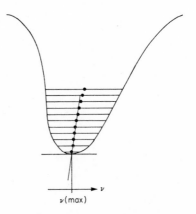

$\nu(max)$

FIG. 5.1. The method of mean intercepts and the location of the maximum of a Band.

change of absorbance with wavenumber is zero at the maximum by definition. A direct reading is very sensitive to noise, distortion of the band by
instrument imperfections, and subjective errors. More reliable and more
valid† is the method of mean intercepts [91]. The investigator measures the
wavenumbers at which the absorbance (or, alternatively, percentage
transmittance) takes a particular value intermediate between the baseline
and absorption peak values, and calculates the mean. This step is repeated
for a series of absorbance values tending towards that at the maximum, and
the wavenumber at the maximum is then found by extrapolation. The
process is shown diagramatically in Fig. 5.1, although, of course, it is not
necessary actually to plot the positions of the mean intercepts. The
subjective error in the reading of the maximum absorbance (or minimum
transmittance) is small, and in any case has little effect on the mean
intercept value.

Correcting for Baseline and Related Effects

The apparent absorbance of a sample is the sum of the baseline
absorbance and the true absorbance. For the reasons given in Chapter 4,

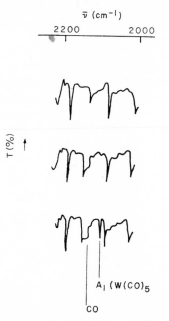

FIG. 5.2. The Highest Frequency peak of $W(CO)_5$ at different concentrations; Hydrocarbon solvent imperfectly compensated.

† Reliable: giving the same result on different occasions. Valid: giving results in accord
with fact.

the baseline is unlikely to be perfectly flat, and baseline spectra should be collected over the entire region of interest. If the spectral trace presents absorbance rather than transmittance, the true absorbance may readily be found by subtracting the baseline, although for most purposes there is no need actually to do this. All that is usually necessary is to check the validity of peaks and to assess the intensities and positions of those due to the sample.

The validity of peaks may be assessed by direct comparison with the baseline. It will usually be possible to say at once which weak peaks are due to solvent or related spurious effects. The danger exists that weak peaks near baseline irregularities may escape detection, and careful comparison of the shapes of peaks at different sample concentrations affords the only protection against this (see e.g. Fig. 5.2).

Where the baseline is flat but deviates from zero absorbance, peak positions are unaffected, and the correction of intensities presents a trivial task. Where the baseline is not horizontal, more care is required. If the spectrum is plotted on a linear absorbance scale, the true intensity at any point is found by direct subtraction of the baseline, the point where the observed peak runs parallel to the baseline gives the peak maximum, and the wavenumber at the maximum can be found by the "mean of intercepts"

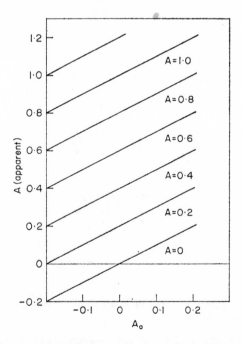

FIG. 5.3. Lines of equal true absorbance against linearly varying absorbance background.

method using lines parallel to the baseline (Fig. 5.3). If the spectrum is plotted on a linear transmittance scale, the true transmittance is given by the ratio of apparent to baseline transmittance. Above a linearly sloping baseline, the lines of equal true absorbance (Fig. 5.3) are replaced by curves of equal true transmittance (Fig. 5.4), so that the apparent position of a peak depends slightly on sample concentration. It is still acceptable to correct for baseline slope and other features by visually estimated subtraction, but the feature subtracted must be scaled down by a facto rof T(total) T(baseline). If baseline features are pronounced and sharp, there may be no reliable way of correcting for them, since the finite optical slit width will vitiate the assumption of simple additivity of absorbances. A related spurious feature of solution spectra that is not eliminated by a baseline correction is "negative solvent absorbance", due to replacement of solvent by solute. This, however, is only liable to be troublesome for concentrated solutions, which in the carbonyl region, at least, are rarely used.

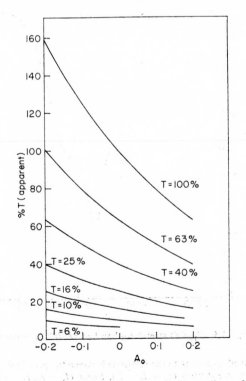

FIG. 5.4. Curves on transmittance scale corresponding to lines of Fig. 5.3.

The Shapes of Peaks

It is very easy, especially for inexperienced workers, to imagine that a band envelope contains more different vibrational peaks than is really the case. A more sophisticated error of the same kind is excessive confidence in peak deconvolution procedures that assume some particular peak shape. In fact, peaks differ considerably in shape even in the same spectrum, so much so that these differences can be of use as an aid to assignment.

Vapour phase spectra the observed form of a vibrational band depends on its assignment, on the symmetry (and in particular on the rotor type†) of the molecule, and on the values of the moments of intertia. For metal carbonyls, the separate rotational transitions will not be resolved but will give rise to splitting of bands in the infrared spectrum into so-called P, Q, and R branches (13, Appendix XVI). For spherical rotors, and for parallel (i.e. z-axis polarised) transitions of symmetric rotors, the central Q branch is rather sharp, so that the three branches are well resolved, as in v_6 of

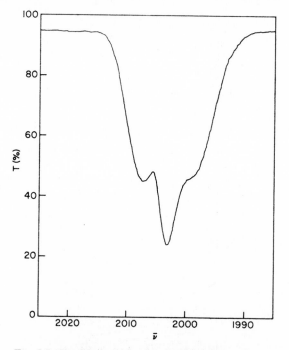

FIG. 5.5. T_{1u} CO stretching band of Mo(CO)$_6$ vapour.

---

† Rotors are classified according to the relations between the principal components of their moments of inertia as *spherical* $(I_x = I_y = I_z)$, *linear* $(I_x = I_y; I_z = 0)$, *symmetric* $(I_x = I_y \neq I_z; I_z \neq 0)$ and *asymmetric* (all three components different).

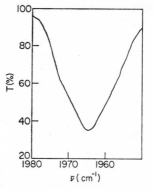

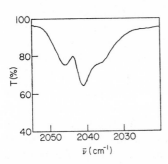

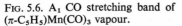

FIG. 5.6. A_1 CO stretching band of (π-C_5H_5)Mn(CO)$_3$ vapour.

FIG. 5.7. E CO stretching band of (π-C_5H_5)MnCO)$_3$ vapour.

Mo(CO)$_6$ and ν_1 of (π-C_5H_5) Mn(CO)$_3$ (Figs 5.5, 5.6). For the perpendicular (x, y polarised) bands of symmetric rotors, such as the E ν(CO) band of (π-C_5H_5) Mn(CO)$_3$, the Q branch is broader, because the selection rules for rotational energy changes are less rigorous, and the resolution of the P and R branches is correspondingly worse (Fig. 5.7). These features can be helpful in assigning the peaks of symmetric or near-symmetric rotors, but make the resolution of weak or closely-spaced vibrational bands of vapours considerably more difficult.

Solids and Mulls. It is dangerous to try to draw detailed conclusions from the spectra of solids. As discussed in Chapter 6, such spectra are not in general simply related to the structure of the isolated molecule. Degenerate bands are frequently split, but so in some situations are non-degenerate bands. A mulled sample may in fact form a finely dispersed solid, or a true solution in the mulling agent, or it may be present in both of these states. It is difficult to tell by mere visual inspection which of these conditions obtains, particularly if the refractive index of the sample material is similar to that of the mulling agent. However, highly irregular peak shapes and a surprisingly rich spectrum tend to indicate the presence of solid. If the spectrum consists of sharp peaks superposed on broader absorptions, the sample material may well be present in both solid and solution phases.

Solution Bands. Bands observed in solution may be formally allowed in the idealised point group of a molecule, or they may become allowed because of internal molecular deviations from ideality, or they may be observable only because of environmental effects.† Bands that are ideally allowed are usually

---

† Workers not prepared for the existence of bands forbidden in the idealised symmetry will feel impelled to assign such bands to impurities and may spend considerable time trying to remove them.

the strongest, and those due to environmental effects weakest, but the three classes overlap. This overlap is only to be expected. The distinctions between ideal and actual symmetry and between intrinsic and environmental distortion in solution, are largely artefacts of description. As well as being weak, bands due to distortions tend to be broad, with ill-defined maxima. This is because of variations between molecules in the size of the distortions that lead to the band being observed. The greater the distortion, the higher the intensity, so that the absorption of those molecules whose frequency differs from the mean is preferentially enhanced.

Among ideally allowed bands, those due to non-degenerate vibrations are usually sharper than those arising from degenerate modes. The reasons for this are as demonstrated in Chapter 3. Small deviations from symmetry, such as are likely to be imposed by fluctuations in solvent environment, cause first order splitting of degenerate modes, so that, summed over the differing environments around each solute molecule, they lead to band broadening. Deviations from idealised symmetry due to the shapes of the non-carbonyl ligands are similar in their effects, and can even lead to slight splitting of the degenerate modes. Non-degenerate bands that arise essentially from the vibration of one carbonyl group are broadened to a smaller extent than are degenerate bands. Non-degenerate, ideally allowed, bands that arise from modes delocalised over the entire molecule are much less broadened. Non-degenerate bands of either kind can only be split if the different rotamers or solvated species are chemically distinct entities, with a bi- or multimodal distribution of environments rather than a continuous spread.

Satellite Bands at Natural Isotopic Abundance

Specimens of carbonyls $M(CO)_n$ contain more than $n\%$ of molecules with one ^{13}CO group. The theory of the spectra of the isotopically labelled species is discussed in Chapter 3. The results of this discussion are relevant to the forms of routinely observed spectra.

Bands due essentially to the vibration of one CO group show weak low-frequency satellites. The theory of Section 3.2 shows that these will be displaced to low energy by around $2\frac{1}{4}\%$ or 40–$45\,cm^{-1}$. Bands due to several CO groups may involve in-phase or out-of-phase coupling. As discussed at several points in Chapter 3, isotopic substitution will lead to the appearance of a well-defined satellite bands to low frequency of the out-of-phase mode, the frequency difference being typically around 35–$40\,cm^{-1}$. There is also a much weaker satellite band a few wavenumbers to the low frequency side of the in-phase mode. This is not generally resolvable, but gives the band arising from the in-phase mode a characteristic skew appearance.

Analytical Expressions for Band Shapes

It should be obvious by this point in the discussion that it is not possible to pre-judge the form of infrared bands. It is nonetheless sometimes convenient to use an analytically defined curve as a simple mathematical model of the actual band envelope. Expressions commonly used for this purpose are the *Gaussian*

$$\varepsilon = \varepsilon_0 \exp a(v - v_0)^2 \tag{5.1}$$

(Fig. 5.8) and *Lorenzian*

$$\varepsilon = a/[(v - v_0)^2 + b^2] \tag{5.2}$$

(Fig. 5.9)†.

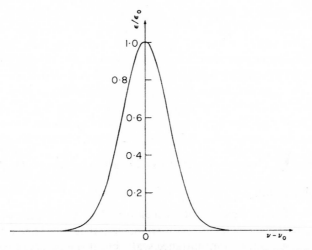

Fɪɢ. 5.8. Gaussion absorption band; height and area match Fig. 5.9.

So far the discussion has ignored the possible effects on band shape of the finite optical slitwidth. This can be expressed by the *slit function* $S(v)$ which affects the observed transmittance at frequency v_0 according to the equation

$$T_{obs}(v_0) = \int_{v = -\infty}^{\infty} S(v) T_{true}(v_0 - v) \, dv \tag{5.3}$$

For a slit of zero width, $S(v)$ is the Kroneker delta function,‡ and the

† The Lorenzian form describes the effect on spectra of random perturbations of duration far less than that of a vibration, while the Gaussian form is a better representation of the effects of a slowly but randomly fluctuating environmental perturbation. Neither model is particularly apt for strongly absorbing species in solution.

‡ This is a function that is nil at all values of the argument except zero, yet has an integrated value of one.

observed shape of a band corresponds to the true shape. If the width of the slit is very much greater than that of the band, the observed band shape is an artefact of the instrument only. Real cases are intermediate, and there is no simple way of deconvoluting the slit and band width effects, though the effects of a finite triangular slit on a Lorenzian band have been analysed [83].

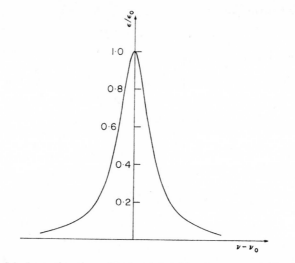

FIG. 5.9. Lorenzian absorption band; height and area match Fig. 5.8.

Overlapping Bands

The forms of overlapping bands depend on how the widths of the bands originate. If the main factor is instrumental, then the bands can best be resolved using a linear transmittance scale, but if the observed band form is due mainly to the sample then a linear absorbance scale is preferable. Whichever scale is chosen, the detection and separation of overlapping peaks is to some extent a subjective process. The form of one component can be estimated by mirror-imaging, curve analysis, or simple extrapolation, although none of these methods are particularly dependable. Mirror-imaging assumes that each isolated band is symmetrical about a vertical line passing through the band maximum. Curve analysis consists in finding the best combination of analytically defined curves to fit the observed spectrum. Simple extrapolation is self-explanatory, and is probably not inferior to the more formal methods; its use in the estimation of the position of a shoulder is illustrated in Fig. 5.10. "Correcting" for the major peak in order to locate the shoulder is formally very similar to correcting for baseline irregularities. Thus, if the observed peak height can be taken as the sum of its components, the maximum in the minor peak occurs where the resultant

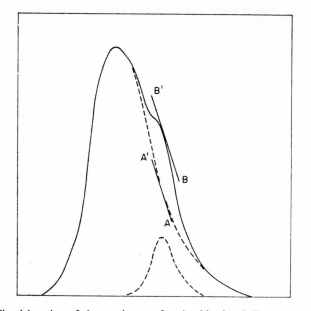

FIG. 5.10. Visual location of the maximum of a shoulder band. Dotted lines represent components. Note the possibility of optical illusions; lines *AA'*, *BB'* need not *look* parallel.

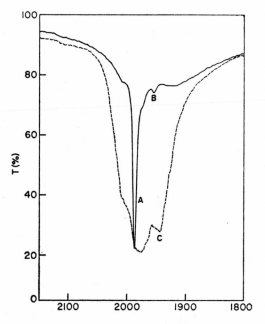

FIG. 5.11. Spectrum of $Mo(CO)_6$ in Nujol mulls of differing thickness.

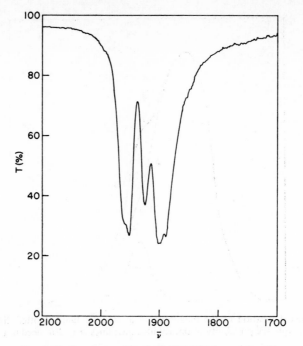

FIG. 5.12. Spectrum of [π-C₅H₅Mo(CO)₃]₂, nujol mull.

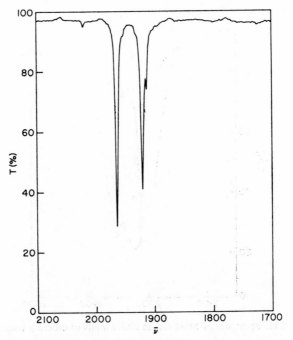

FIG. 5.13. Spectrum of [π-C₅H₅Mo(CO)₃]₂, cyclohexane solution.

curve runs parallel to the major peak curve. There is no guarantee in any case that peak heights are truly additive, and it may be helpful to collect spectra at several different sample concentrations in order to compare the appearances of "shoulder" peaks.

5.2. Discussion and Assignment of Some Typical Spectra

In this section a few typical spectra are discussed in some detail so as to illustrate the general points raised above. It will become apparent that the discussion of any one spectrum relies implicitly on comparison with a large number of others, and detailed comparison with results already known constitutes an essential test of any proposed assignment.

The Effects of Phase

Figure 5.11 shows spectra of $Mo(CO)_6$ recorded in a very thick nujol mull, and of the same mull thinned out by addition of more nujol. The first spectrum is mainly that of the solid. The band observed is very broad, and split into at least three, and probably more, distinct components. The second spectrum is clearly qualitatively different. The sharp peak A that dominates the spectrum must be due to material in solution ($Mo(CO)_6$ is appreciably soluble in hydrocarbons), while peak B is probably, from its position, an isotope satellite band. The shoulder C in the spectrum of the more concentrated mull is then also assignable to $Mo(^{12}CO)_5$ (^{13}CO), and it is evident that the solid is non-cubic and that intermolecular interactions are considerable.

The spectra of $[\pi\text{-}C_5H_5Mo(CO)_3]_2$ in a nujol mull and in hydrocarbon solution are shown in Figs 5.12, 5.13. In this case, the solubility of the sample material in nujol is low, so that the mull spectrum is that of a solid. The difference between the mull and solution spectra is as great as if they came from different compounds.

Effects of Concentration and Solvent

The spectra of $[\pi\text{-}C_5H_5)Mo(CO)_3]_2$ in saturated and dilute solution in methylene chloride are shown in Fig. 5.15. There are several features of interest.

Altering concentration alters the appearances of the spectrum on a linear transmittance scale. Peaks B and C appear almost to have the same maximum intensity at high concentrations, while in dilute solution the difference is marked and indicates a factor of 3:2 between the extinction coefficients at the two maxima.

The high energy band A is much weaker, and correlates with an exceedingly weak band in the spectrum in hydrocarbon (Fig. 5.13.) The

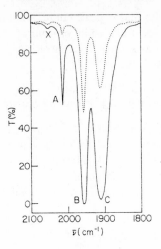

FIG. 5.14. Spectrum of $[(\pi\text{-}C_5H_5)Mn(CO)_3]_2$ in CH_2Cl_2.

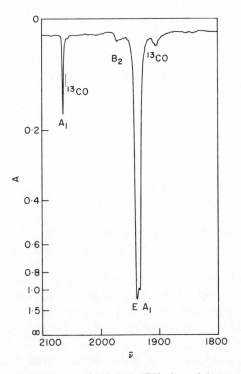

FIG. 5.15. Spectrum of $(cx)_3PMo(CO)_5$ in cyclohexane.

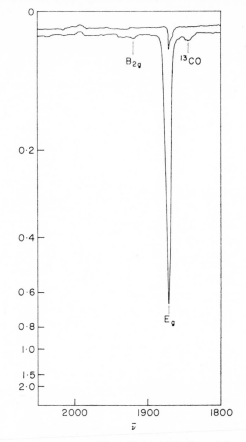

FIG. 5.16. Spectrum of $[(cx)_3P]_2Mo(CO)_4$ in cyclohexane.

form of band A is best studied in concentrated solution, where the expected skewing to lower wavenumber becomes apparent. Band C is a broad composite containing bands resolved in hydrocarbon solutions. The band X is background, and is identified as such by its failure to grow with concentration.

Some Examples of Solution Bands

The solution spectrum of $(cx)_3 PMo(CO)_5$ (cx = cyclohexyl) is shown in Fig. 5.16. The molecule belongs to case (iv) of $M(CO)_5$, discussed in Section 3.6. The highest frequency band (I) is due to the in-phase motion of four CO groups. It is sharp, as predicted, and shows the characteristic skew to low wavenumbers. Band II is due to the B_2 mode of the equatorial CO groups, and acquires its intensity from their distortions from planarity.

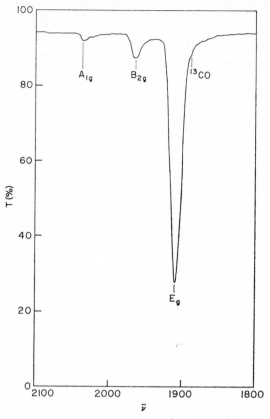

FIG. 5.17. Spectrum of *trans*-[(Pr^{i}O)$_3$P]$_2$Mo(CO)$_4$.

Bands III and IV are the predicted E and $(A_1)_1$ bands, while V is the low-frequency isotope band associated with the E mode. No separate isotope band can be seen for the $(A_1)_1$ mode. The spectrum of the related trans-disubstituted species is shown in Fig. 5.16. The $(A_1)_1$ mode has vanished with the disappearance of the axial CO group. The isotope satellite of the E band is still observable, at some 30 cm^{-1} to low frequency. The B_2 mode is just visible. Finally, the spectrum of *trans*[(Pr^{i}O)$_3$ P]$_2$ Mo(CO)$_4$ shows pronounced effects due to irregular packing of the PrO fragments. The allowed (E_u) band is severely broadened, so much so that the ^{13}CO satellite is no longer resolved, and not only the B_{2g} but the A_{1g} mode as well gives rise to weak, severely broadened bands. These are distinguished from impurity bands by their constancy from one specimen to another, and by analogy with the corresponding modes detected by infrared or Raman spectroscopy of related molecules.

Chapter 6

Some Topics Related to the Chemical Interpretation of Vibrational Data

This chapter is concerned with a number of topics that are relevant to the chemical interpretation of vibrational data. A survey of the principles used in the analysis of combination spectra introduces a discussion of the importance of anharmonicity, and of the validity of any force constant or parameter analysis in which anharmonicity is ignored. The assignment of bands other than those due to CO stretching modes is described, and their possible use in informal treatments of metal–carbon bond strengths evaluated. The intensity of the carbonyl infrared bands is considered in relation to orbital following effects, which are shown to have some influence on force constants. Finally, the analysis of the spectra of solids is described, and it is shown how such spectra can be valuable in the detailed assignment of solution bands, as well as being of interest in their own right.

6.1. Combination and Overtone Spectra

It was stated in Section 2.2 that the observed frequencies of a molecule differ from the mechanical frequencies because of anharmonic terms in the potential energy. The gravity of these effects can be estimated from the combination and overtone bands, which are observable in the infrared spectrum (Section 4.4). The symmetry species to which a combination band belongs can be found by direct multiplication of the symmetry species of the fundamental components, so that for a binary combination involving the excitation of modes a, b,

$$\Gamma \left(v_a = 1, v_b = 1\right) = \Gamma_a \times \Gamma_b. \tag{6.1}$$

The symmetry species of overtone bands are those of the *symmetric products*† of the fundamentals, so that for a binary overtone of mode a

$$\Gamma\,(v_a = 2) = [\Gamma_a]_2^+ \qquad (6.2)$$

where the character of $[\Gamma_a]_2^+$ under any operation R is given by

$$\chi_2(R) = \{\chi_1(R^2) + [\chi_1(R)]^2\}/2. \qquad (6.3)$$

The symmetries of higher overtones may be found from the equations

$$\chi_n(R) = \{\chi_1(R)\,\chi_{n-1}(R) + \chi_1(R'')\}/2 \qquad (6.4)$$

for doubly degenerate representations, and

$$\chi_n(R) = \{4\chi_1(R)\,\chi_{n-1}(R) + [\chi_1(R^2)$$
$$- \chi_1(R)^2]\,\chi_{n-2}(R) + 2\chi(R'')\}/6 \qquad (6.5)$$

for triply degenerate representations. The symmetries of binary and ternary combination bands in some common point groups are listed in Appendix III.

The approximate energies of combination and overtone bands are given by Eqs (2.11, 2.12). For degenerate modes, the summations in these equations should be taken over all the individual components‡, or else supplemented, as described in the footnote to Eq. (2.12), to give

$$E/hc = \sum_i \omega_i(v_i + g_i/2) + \sum_{i<j} \sum_j X_{ij}(v_i + g_i/2)\,(v_j + g_j/2) + \sum_i G_{ii}\,l_i^2$$
$$+ \text{ other terms in } G_{ij}\,l_i\,l_j \qquad (6.6)$$

or

$$E/hc = \sum_i X_i'\,v_i + \sum_{i<j} \sum_j X_{ij}'\,v_i\,v_j + \sum_i G_{ii}\,l_i^2$$
$$+ \text{ other terms} \qquad (6.7)$$

(The terms in G_{ij} are as yet unused for metal carbonyls; g_i is the degeneracy of the ith mode.)

The importance of anharmonicity must be decided by experiment. The first step in the investigation of any particular compound is obviously

---

† This is because the total wavefunction is symmetric under interchange of the labels of degenerate quanta of vibrational energy. Contrast the antisymmetric product that gives the triplet terms of a doubly occupied degenerate electronic energy level, where the wavefunction changes sign on interchange of the electron labels.

‡ There are obvious symmetry restrictions. Thus if the modes i, j, k span a three-fold degenerate representation.

$$X_{ii} = X_{jk} = X_{jk};\ X_{ij} = X_{jk} = X_{ki}.$$

to assign its combination spectrum. This can often be performed by inspection, finding assignments that give rise to acceptable deviations between the combination frequencies and the sums of their components. It is commonly found that the highest carbonyl mode, which always belongs to the fully symmetric representation, combines with all the infrared active modes. In such cases, the highest frequency band observed is one of a family with very nearly the same relative spacing as the fundamental infrared spectrum. Some bands may be extremely weak, and it may sometimes prove possible to foresee this weakness and use it in assignment. For example, the $2e$ combination of $Mn(CO)_5Br$ (C_{4v}) is expected to be weak since it correlates with a forbidden $2t_{1u}$ mode of octahedral $M(CO)_6$, and this prediction is indeed borne out [92].

The Effects of Anharmonicity

Once the full spectrum of any species is assigned, the anharmonicity constants and harmonic frequencies of Eqs (2.11, 2.12) may in principle be calculated. In practice a large number of problems arise. Many combinations and overtones are infrared inactive. They are thus inaccessible, since carbonyl combination bands are too weak to be observed in the Raman spectra of solutions.† Fundamental and binary data are insufficient to determine the anharmonic constants. In some cases ternary spectra give enough additional information to do this, but ternary bands are weak, and have so far only been described for a very few species [33, 94]. Finding all the constants commonly involves resolving overtone bands of degenerate modes into their components. The bands are, in some cases at least, too broad for such resolution to be possible. For example, the $3T_{1u}$ bands of $Cr(CO)_6$ and $Mo(CO)_6$ (in CCl_4) are not clearly split, although the expected splitting is observable in the comparable band of $W(CO)_6$ (dissolved in CS_2). Transfer of anharmonic constants between compounds may help with this problem, but is clearly only possible for very closely related compounds.

There are also difficulties of interpretation. Anharmonicity is generally taken to arise from cubic and quartic terms in the molecular potential energy. The full potential energy expression of any molecule must, of course, belong to the fully symmetric representation of the molecular point group, and this condition generally places restrictions on the number of independent parameters in each part of the force field. Nonetheless, the number of parameters exceeds the number of independent constants in Eqs (2.11, 2.12), and so these equations impose arbitrary concealed constraints on the force field.‡ In particular, it is assumed that interactions between different

† This statement is of course liable to be overtaken by events; but it is relevant that the $2T_{1u}$ overtone bands were not observed in a spectrum of liquid $Mo(CO)_6$ [93].

‡ The author thanks Dr. J. R. Miller for drawing his attention to this fact.

combination or overtone species may be ignored; but there are no grounds for this assumption, since the bands in question are close together in frequency and, when of the same symmetry, satisfy the conditions for Fermi resonance.

It may also be possible for binary and more especially ternary bands to be distorted by solvent interaction in ways that have far less effect upon fundamentals; and, in fact, these bands are observed to be quite broad. The solvent field could simply be regarded as part of the potential field that gives rise to the anharmonicity constants, but for the possibility that the different band maxima refer to different molecular populations. This possibility arises from the nature of the processes that render combination and overtone bands observable. For a pure harmonic oscillator, giving rise to an oscillating electric dipole linearly dependent on displacement, the perturbation produced by electric dipole radiation (to which light radiation effectively approximates) is linear in displacement. As a result, the vibrational quantum number can change only by one unit on absorption. Cubic and quartic terms have the effect of distorting the form of the wavefunctions, so that combinations and overtones acquire some dipole allowed character. In addition, non-linearity in the oscillating electric dipole of the molecule gives rise to non-linear terms in the effect of light, so that dipole radiation can give rise to quadrupole allowed transitions. Both of these effects can be modified by solvent interactions. If the effect of solvent on intensity could be ignored, the band maximum would correspond to the statistically most favoured molecular configuration or environment. Even where solvent effects on position are large enough to cause appreciable broadening, this interpretation of a maximum remains valid. Unfortunately, the near-forbidden nature of binary and ternary bands suggests the possibility that irregular solvent packing can considerably enhance absorption in these regions by some particular configurations; the effect is similar to that noted in Chapter 5 for ideally forbidden fundamentals. This may perhaps cause the band maximum to deviate from the position most favoured on purely statistical grounds, and the deviations could be different for different non-fundamental bands.

Despite these difficulties of interpretation, Eqs (2.11), (2.12) can be used in a purely descriptive manner to compare observed spectra, and if the parameters are neither over- nor under-determined, then they will be operationally defined. Unfortunately, like all purely operational quantities, they may differ from the quantities of theoretical interest; thus the ω defined by fitting data to Eq. (2.11) may or may not be equal to the desired mechanical frequencies.

The criticisms levelled here at the best available estimates of anharmonicity constants and mechanical frequencies should not be taken to mean that anharmonicity is unimportant. On the contrary, it is precisely because anharmonicity may be very important indeed that the author feels the need to argue that our knowledge of its affects is less than definitive. The size and nature of the effects involved can be estimated from those studies that enable the ω of Eq. (2.11) to be calculated. It is then possible to compare force constants and parameters calculated from the best estimated harmonic frequencies with the corresponding "anharmonic" quantities calculated from the observed frequencies. Four questions about the consequences of anharmonicity require an answer. Does the use of mechanical frequencies lead to a significantly changed estimate of individual stretching force constants or parameters? Are all modes affected in much the same way, or do differences arise, leading to revised values for interaction parameters? What changes occur in the relative values of parameters for different kinds of carbonyl group in the same molecule? And finally, how are comparisons between different molecules affected?

TABLE 6.1. The effects of anharmonicity in $Mo(CO)_6$ and $Mn(CO)_5$ Br (a).

Species	Quantity	Uncorrected value	Harmonic value	Reference
$Mo(CO)_6$	v_1	2116·7	2140·2	[23]
(in CCl_4)	v_3	2018·8	2037·3	
	v_6	1986·1	2026·2	
	$F(CO)$	1665	1715	
	$F(CO, C'O')_c$	16	17	
	$F(CO, C'O')_t$	20	1	
$Mn(CO)_5Br$	$v(A_1)$	2135·2	2156·1	[33]
(in CCl_4)	$v(B_2)$	2083	2087·7	
	$v(E)$	2052·1	2085·7	
	$v(A_1)$	2001·5	2025·0	
	$k(a)$ (b)	1635	1679	
	$k(e)$	1741	1784	
	c	18·6	25	
	d	30·5	37	
	t	43·2	24	
	$F(a)$	1627	1671	
	$F(e)$	1733	1776	
	$F(c)$	7	13	
	$F(d)$	19	25	
	$F(t)$	34	15	

(a) Frequencies in cm^{-1}, force constants in Nm^{-1}.
(b) Force constants derived from the energy factored force field of [299], using the estimates of [33] for the effects of coupling between CO and other modes, and of anharmonicity. $F(a)$, $F(e)$, $F(c)$ etc. are estimated force constants corresponding to the parameters $k(a)$, $k(e)$, etc.

A partial answer to all these questions can be found by examining the implications of three detailed published studies. The first two of these refer to the Group VI metal hexacarbonyls [94], and to pentacarbonylmanganese bromide [33]. In both these studies, use was made of ternary and binary infrared spectra as well as of fundamental Raman spectra to solve Eq. (6.6). Both these studies also contain values (quite rigorous in the case of the hexacarbonyls, slightly less so for the pentacarbonyl bromide) of the harmonic and anharmonic values of the carbonyl force constants and parameters.

The results are shown in Table 6.1. It is obvious that anharmonicity has a marked effect both on the observed frequencies and on the force constants and parameters calculated from these; indeed, the stretching constants and parameters are all shifted by approximately 50 Nm^{-1}. The different vibrational modes are affected to different degrees, so that the relative values of the interaction constants are sharply altered. The *cis* interaction force constants remain almost unchanged, but the *trans* constants are seriously reduced; the same effect is found for the interaction parameters. $Mn(CO)_5Br$ contains two physically different carbonyl groups. It is found that the difference between their stretching constants (or parameters) is altered only by 1 Nm^{-1}, which is less than the experimental uncertainty. The relative values for $Cr(CO)_6$ and $Mn(CO)_5Br$ are hardly altered.

The third relevant study [95] concerns a range of species of the type $(\pi\text{-}C_5H_5)Fe(CO)_2X$. Since the CO groups in these molecules are not collinear all the fundamental and binary transitions are infrared active, and

TABLE 6.2.

Species	Quantity	Uncorrected value	Corrected value
$CpFe(CO)_2Cl$ (a, b)	$v(A')$	2051·4	2073·8
	$v(A'')$	2007·8	2033·5
	k	1664	1703
	i	36	33
$CpFe(CO)_2Br$	$v(A')$	2046·5	2073·8
	$v(A'')$	2004·0	2036·3
	k	1657	1706
	i	35	31
$CpFe(CO)_2I$	$v(A')$	2038·8	2064·2
	$v(A'')$	1997·7	2027·9
	k	1646	1691
	i	34	30
$CpCo(CO)_2$ (c)	$v(A')$	2025·3	2046·7

(a) Data for this and the following species refer to solutions in CS_2.
(b) Results for a range of species of type $CpFe(CO)_2MR_3$ are broadly similar.
(c) $2v\ (A)''$ obscured by ring absorption.

give enough information for the quantities of Eq. (2.11) to be evaluated. While the authors prudently refrain from claiming as much, the quantities estimated include the mechanical frequencies, from which it is possible to derive the harmonic values of the stretching parameters. The quantities calculated are compared in Table 6.2 with the corresponding anharmonic parameters. As in the studies described earlier, the anharmonicity correction increases the value for the stretching parameter by around $40 \, Nm^{-1}$. The effects on the interaction parameters are small, resembling (as might be expected) the effects noted for mutually *cis* CO groups in the hexacarbonyls. There is only one result that runs contrary to chemical expectation; the harmonic parameter found for $(\pi\text{-}C_5H_5)Fe(CO)_2Br$ is higher than that for the chloride. The difference is admittedly small; but it is disturbing that the *uncorrected* (anharmonic) parameters show the bromide in its expected position, midway between the chloride and the iodide.

It is now possible to give tentative answers to the questions raised. The use of mechanical frequencies leads to stretching parameters some 40 to $50 \, Nm^{-1}$ higher than those found from the observed frequencies; this is a very considerable absolute difference. Different modes involving the motion of mutually *trans* carbonyl groups are differently affected, being brought closer together by the correction, so that the *trans* interaction parameter is reduced; but *cis* interaction parameters are only marginally affected. Different carbonyl groups in the same molecule are affected to almost the same extent, as are all carbonyl groups belonging to very closely related series of molecules, but there are some slight differences (of around $10 \, Nm^{-1}$) between different molecular series.

To summarise, correcting for anharmonicity increases stretching parameters and constants, but on the whole reduces interaction parameters, especially between mutually trans CO groups. There seems little advantage in the use of harmonic parameters for general comparisons. Harmonic parameters are, however, essential for any kind of absolute estimate of the mechanical force constants at equilibrium. Less apparent are the relative merits of harmonic and anharmonic parameters in the analysis of the forms of the normal modes (as in studies of intensity). The harmonic values only describe the molecule correctly near equilibrium. It could therefore be argued that the anharmonic values lead to a better estimate of the degree of mixing of symmetry coordinates that takes place over the amplitude of an excited vibration.

The Location of Inactive Modes from Combination Spectra

Very often, a band is infrared inactive in the fundamental region but forms a component of allowed binary bands. It is then possible to estimate the inactive fundamental frequency by analysis of the binary spectrum, provided the effects of anharmonicity can be ignored. Generally speaking, they cannot, and an observed combination or overtone band is less than the sum of its components. Thus if v_a is inactive, but v_a and the combination $(v_a + v_b)$ are active, the quantity $(v_a + v_b) - v_b$ is a lower limit, rather than a reliable estimate, for v_a.

Sometimes such a lower limit is all that is required. For instance, if the highest observed binary band occurs at more than twice the frequency v_f of the highest observed fundamental, it follows at once that there is at least one (unobserved, and possibly inactive) fundamental frequency higher than v_f[59]. In comparisons of closely related series of compounds it seems plausible that errors due to non-additivity will not invalidate trends, though they may well cause the trends to be underestimated.† The actual size of the errors due to non-additivity can be estimated when v_a, v_b and $(v_a + v_b)$ are all known; some specimen values are shown in Table 6.3. It is evident that for some compounds non-additivity is a serious problem, while for others it is not. The values for polynuclear species appear to be better than those for compounds containing one metal atom only, and the potential error due to non-additivity is least for those compounds in which the vibrations are dispersed over the largest number of carbonyl groups with the weakest mutual interactions.

Provided that the highly variable quality of estimates of infrared inactive frequencies from combination data is borne in mind, such estimates do serve a useful purpose. They may be obtained for coloured or photosensitive compounds, using instrumentation that is widely available; they have provided semiquantitative evidence of the importance of the interactions between sub-units of polynuclear species; and on occasion they can lead to values for frequencies that are inactive in the Raman as well as in the infrared spectrum. Thus some of the earliest evidence of interaction between the local modes of the two moieties of $Mn_2(CO)_{10}$ came from combination studies [96, 97], and combination studies also led to the assignment of a regular tetrahedral structure to the species $[Mn(CO)_3SR]_x$, and to the assignment and an estimate of the frequency of the totally inactive T_1 carbonyl mode in such species [31, 32].

Location by Combination Spectroscopy of Modes other than CO Stretching Modes

A brief word should be said about the location of modes due to processes other than CO stretching. Combinations involving such modes, or involving

† For example, the quantity of interest may be the difference between the highest vibrational frequency and the highest frequency that is infrared active. This difference generally arises as a result of interaction parameters. Larger interaction parameters are probably associated, in otherwise comparable situations, with larger cross-anharmonicity constants, leading to a more serious underestimate of the quantity of interest.

one such mode and a $v(CO)$ mode, have been analysed for the hexacarbonyls [23, 94]. Where both frequencies are known, the non-additivity error of the combination band is generally less than 3 cm^{-1}, as against 9 and 20 cm^{-1} for the $A_{1g} + T_{1u}$ and $E_g + T_{1u}$ combinations of CO modes with each other. Thus for reasons that are not totally obvious (the amplitudes of metal–carbon vibrations, for example, are greater than those of CO stretches, as are the harmonic cross-terms connecting the individual bond distortions) the non-additivity error for $v(MC)$, $\delta(MCO)$, and $\delta(CMC)$ modes may be ignored.

6.2. Assignment and Interpretation of Modes Other than $v(CO)$

In addition to the CO stretching bands, all metal carbonyls possess frequencies in the region 300–700 cm^{-1}, due to M–C stretching and M–C–O deformation modes. Other bands in the spectrum are those due to CMC deformation (in the region 50–120 cm^{-1}), to the internal vibrations of other ligands, to other metal–ligand vibrations and to L–M–C distortions (where L is a ligand other than CO). CMC and LMC deformations occur at low frequencies and will not be discussed further in this section. Internal vibrations of other ligands may be recognised from the pure ligand spectra, and M–L stretching and L–M–L bending vibrations lie strictly outside the scope of this book; they may on occasion be recognised from the spectra of model compounds not containing CO. $v(MH)$ and $v(CO)$ motions can in some circumstances couple. This situation may be recognised by the sensitivity of "$v(CO)$" to deuteration [98].

The remaining bands can be assigned to their symmetry classes, at least in carbonyl complexes with a high degree of symmetry, using a range of arguments, such as infrared and Raman activity, Raman depolarisation, and combination band infrared activity. The number of CO, MCO, MC and CMC vibrations of each class can be predicted from simple group-theoretical arguments, and the separation of high, medium, and low frequency regions is clear.

The only ambiguity can arise in the middle frequency region, within which the MCO and MC frequencies commonly overlap. If there are MCO and MC frequencies of the same symmetry class, they will interact, so that the actual modes will contain contributions from both. Nonetheless, it is still important in exact studies to decide which mode derives predominantly from which kind of distortion. A wrong assignment will lead to an incorrect choice between the possible solutions to the secular equation for the symmetry block of interest, and hence to erroneous values for the force constants.

Several methods are available for distinguishing between MC and MCO modes of the same symmetry class. The most reliable is isotopic substitution. It was shown in Section 3.2 that MCO bands are particularly

TABLE 6.3. Anharmonicity errors in combination spectra (a)

Species	ν_a	ν_b	$(\nu_a) + (\nu_b)$	$(\nu_a + \nu_b)$	$(\nu_a) + (\nu_b) - (\nu_a + \nu_b)$
$Cr(CO)_6(b)$	$\nu_1 = 2113$	$\nu_6 = 1985$	4098	4089	9
	$\nu_3 = 2019$	$\nu_6 = 1985$	4004	3984	20
$Mo(CO)_6(b)$	$\nu_1 = 2117$	$\nu_6 = 1985$	4102	4095	7
	$\nu_3 = 2019$	$\nu_6 = 1985$	4004	3987	17
$W(CO)_6(b)$	$\nu_1 = 2121$	$\nu_6 = 1980$	4101	4091	10
	$\nu_3 = 2013$	$\nu_6 = 1980$	3993	3974	19
$Mn(CO)_5Br(c)$		$\nu_1 = 2135 \cdot 2$	$2(\nu_1) = 4270 \cdot 4$	$4265 \cdot 0$	$5 \cdot 4$
	$\nu_2 = 2001 \cdot 5$	$\nu_1 = 2135 \cdot 2$	$4136 \cdot 7$	$4132 \cdot 1$	$4 \cdot 6$
	$\nu_{15} = 2052 \cdot 1$	$\nu_1 = 2135 \cdot 2$	$4187 \cdot 3$	$4179 \cdot 1$	$8 \cdot 2$
			$2(\nu_2) = 4003 \cdot 0$	$3982 \cdot 2$	$17 \cdot 8$
			$(\nu_2) + (\nu_{15}) = 4053 \cdot 6$	$4053 \cdot 2$	$0 \cdot 4$
$CpFe(CO)_2Cl(d, e)$		$\nu(A') = 2051 \cdot 4$	$2[\nu(A')] = 4102 \cdot 8$	$4092 \cdot 5$	$10 \cdot 3$
		$\nu(A'') = 2007 \cdot 8$	$2[\nu(A'')] = 4015 \cdot 6$	$4002 \cdot 0$	$13 \cdot 6$
			$[\nu(A')]+[\nu(A'')]=4059 \cdot 2$	$4035 \cdot 0$	$24 \cdot 2$

$CpCo(CO)_2$ (d,f)			
$\nu(A') = 2025\cdot3$	$2[\nu(A')]\ 4050\cdot6$	$4041\cdot0$	$9\cdot6$
$\nu(A'') = 1963\cdot3$	$[\nu(A') + \nu(A'')] = 3988\cdot6$	$3965\cdot0$	$23\cdot6$
$[Mn(CO)_3SEt]_4$ (e,g,h)			
$\nu(A_1) = 2044$	$\nu(A_1) + \nu(T_2) = 4060$	4030	(i)
$\nu(T_2) = 2016$	$2\nu(T_2) = 4032$	3986	2
$\nu(E) = 1954$	$\nu(A_1) + \nu(T_2) = 3986$	3958	0
$\nu(T_2) = 1942$	$\nu(T_2) + \nu(T_2) = 3958$	3935	0
$\nu(T_1) = 1923$	$\nu(T_2) + \nu(T_1) = 3939$		4
	$\nu(E) + \nu(T_2) = 3896$		(i)
	$2\nu(T_2) = 3884$	3886	-2
	$\nu(T_2) + (T_1) = 3865$	3865	0
	$2\nu(T_1) = 3846$		(i)

(a) Frequencies in cm^{-1}.
(b) in CCl$_4$ [94].
(c) in CCl$_4$ [33].
(d) in CS$_2$ [95].
(e) Results for related species are similar.
(f) 2 (A'') obscured by ring absorption.
(g) Frequencies inferred from combination spectrum given in italics.
(h) In CCl$_4$ [31].
(i) Used to evaluate fundamentals; assumed zero.

sensitive to ^{13}CO substitution, which lowers their frequencies more markedly than does substitution with $C^{18}O$; for MC stretching, the reverse is true. Given a number of assignments based on isotopic data or otherwise made free of ambiguity, it may be possible to assign frequencies in related compounds using *filation curves*, or plots of analogous frequencies against degree of substitution by a given ligand. For example, the MC and MCO frequencies of class T_2 in nickel tetracarbonyl have been assigned unambiguously from isotope shifts [99]. The MC and MCO modes in species L_3NiCO belong to the symmetry types A_1 and E, and may be distinguished by Raman depolarisation data. The MC and MCO modes of $LNi(CO)_3$ both span the representation E. An unambiguous assignment of the E modes is nevertheless possible by intrapolation [100]. MC bands are generally more sensitive than MCO bands to substitution of one ligand by another. Replacement by a weaker-accepting ligand generally leads to an increase in $v(MC)$, but has far less effect on $\delta(MCO)$. These substituent effects sometimes make possible an assignment, in the absence of a filation curve, from comparisons of the spectra of related complexes containing different non-carbonyl ligands. The intensity of a single band is not a totally reliable guide to its assignment (though MC modes are generally stronger in the Raman spectrum). It is, however, sometimes possible in complexes of fairly low symmetry to draw useful conclusions from the relative intensities of groups of bands. For example, the oscillating dipole associated with an MC stretch is more or less collinear with the bond. The moment associated with an MCO bend is thought to be at right angles to the group. When more than one CO group is present, the moments due to the deformations of the separate MCO units may be combined by vector addition, so that bands can be arranged in families of calculable relative intensity [101]. For example, in an $M(CO)_2$ unit, if the CMC angle is obtuse, the in-phase CMC mode should be weaker in the infrared spectrum than the out-of-phase mode, but the opposite is true for the in-phase MCO deformations. Whatever the CMC angle (provided it is considerably smaller than 180°), the in-phase out-of-plane bending mode will be the strongest of the bending modes, while the out-of-plane out-of-phase mode, which is inactive in C_{2v}, will be the weakest.

The bonding interpretation of the bending force constants and frequencies is not clear. The metal–carbon stretching constant is expected to increase both with greater carbon–metal bonding, and with greater metal–carbon π-back-bonding. The second of these effects is generally the most important, as the data cited in Chapter 7 show. Less electro-negative ligands generally lead to higher metal–carbon frequencies, even though they presumably cause lowering of the degree of σ-bonding. There seems to be a need for careful comparisons of trends in metal–carbon and carbon–oxygen stretching frequencies. For example, the lower A_1 frequency of species $M(CO)_5L$ arises mainly from the vibration of the CO group trans to L, and is lower when L is an amine than when L is an alkylphosphine. This effect could be due to a trans σ-influence, as discussed in Section 6.4 below, if the amine is acting as a better σ-donor. Alternatively, it could be argued that the phosphine is acting as a π-acceptor. If the difference in frequencies is due to π-effects only, the *trans* metal–carbon stretch should (ignoring coupling to other modes) be lower in the phosphine complex, while if it is due to σ-effects only, the reverse should be true.

One surprising, and somewhat disappointing, result emerges from a comparison of nickel carbonyl with the hexacarbonyls of the chromium group, although the trend (W > Cr > Mo) within the group is satisfactory. It would be naive to expect the spectroscopic data to predict the high chemical reactivity of $Ni(CO)_4$ which is due to a special effect. The activation energy for unimolecular loss of CO from $Ni(CO)_4$ is unusually low [102], but this probably is because of reorganisation within the $Ni(CO)_3$ fragment, which is expected to increase the strength of the remaining bonds [103]. The thermodynamic data are, however, more disturbing. The metal–carbon force constants for $Ni(CO)_4$ and $Cr(CO)_6$ are almost identical. On the other hand, the average chromium–carbon bond dissociation energy, calculated with respect to the "valence state" of the chromium atom and the ground state of carbon monoxide, is 241 kJ mole^{-1}, while the corresponding quantity for nickel carbonyl is 191 kJ mole^{-1}. The thermodynamic and spectroscopic results are not necessarily in conflict, and can readily be reconciled by such *ad hoc* explanations as stronger steric compression round chromium;† but the comparison shows the dangers of trying to relate force constants directly to chemical or even to thermochemical results.

6.3. The Interpretation of Carbonyl Intensity Data

It was shown in Chapter 3 that, when a complex contains only one kind of CO ligand, the intensity distribution of the observed carbonyl stretching modes may be explained by geometric effects alone. In complexes with more than one kind of CO group, both geometric and coupling effects are important as are differences in bond dipole derivative; but provided a force field may be assumed, the various factors may be separated and the relevant angles and relative bond dipole derivatives determined. Calculations of this kind may

† Thus each carbonyl group around chromium experiences a *short range* repulsive force due to the presence of all the others. This will lower the force constant at equilibrium more markedly than the bond dissociation energy. It should in fairness be pointed out that the known shapes of pentacarbonyl halides [104, 105] and the probable shape of matrix-isolated chromium pentacarbonyl [44], point strongly to the reality of some such steric repulsion.

be criticised in detail for possible errors in the form of the normal modes, for considering only the linear terms in the variation of molecular dipole with CO stretch, and for equating the angle between bond oscillating dipoles with the angle between the groups that give rise to them.

Despite such significant matters of detail, the overall picture is quite clear. Carbon monoxide itself does not absorb strongly. When absorbed onto such non-π-donating substrates as metal oxides, its specific absorption changes, as does the CO frequency. Adsorption is accompanied by an increase in CO stretching frequency, which is attributable to σ-donation by CO 5σ to the substrate. The observed intensity changes may be correlated with the frequency changes and hence with the presumed degree of σ-donation. It appears from the results that slight σ-donation causes a reduction in the oscillating dipole of CO, while a greater degree of donation causes a change in sign. Facts of this kind do not explain the observed very high intensities of the CO bands of metal–carbon complexes, which increase along isoelectronic series, with falling CO stretching frequency. These intensities must be a consequence of the metal–carbonyl π-bonding.

The infrared intensities of bands are proportionate to the square of the oscillating dipoles associated with the vibrations concerned. Metal–carbonyl π-bonding provides a mechanism by which this oscillating dipole may be increased. The stretching of a CO group leads to a lowering in the energy of the CO 2π orbital [19], since the antibonding interaction between carbon and oxygen is weakened and this orbital therefore becomes a better acceptor. At the same time, a CO stretching vibration causes a decrease in the metal–carbon distance (Section 3.2), and the overlap between carbon $2p\pi$ (the main component of CO 2π) and the metal d-orbitals in thereby increased. Thus charge may be expected to flow from metal to carbon, so that the observed high oscillating dipole associated with the CO stretch is in fact due to orbital following in the metal–carbon bond. Bridging CO groups (in $Co_4(CO)_{12}$, for example) show lower specific intensities (i.e. intensities per carbonyl group) than do terminal groups in the same complex [106]; this finding accords well with the orbital following theory, since the dipoles that arise from charge flow between the bridged metal atoms and the bridging CO group will to some extent cancel.

Orbital following effects are presumed to be responsible for the interaction constant connecting the MC and CO modes of an individual MCO group. It is therefore to be expected that this interaction constant should correlate with specific intensity. Unfortunately, values of the interaction constant as such are available only for a very few compounds. The *trans* interaction parameters of the energy factored force field are more complex quantities, but it has been argued [107] that nonetheless they are related to the relevant interaction. It is therefore reasonable to seek a general correlation between interaction parameter and average specific absorbance, and such a correlation is indeed found [108].† The absorbance of $Ni(CO)[P(OEt)_3]_3$ is comparable with that of the other nickel–carbonyl complexes considered, confirming that the relevant interactions are within the individual MCO units, and that coupling between the units is not important as such.

Where more than one type of CO group is present in a molecule, the possibility arises of significant differences in specific intensity between these different groups. Such effects are difficult to demonstrate conclusively, and even more difficult to interpret. For example, the differences found between the specific intensities of axial and equatorial CO in a series of monosubstituted molybdenum carbonyls were not significant [109], but major differences were found in pentatcarbonyl manganese halides [40]. There were minor discrepancies of computational procedure between these two studies, but the differences in findings seem too great to have arisen in this way. It could be that special factors are involved, such as orbital following in the antibonding π^*-system formed by halogen and metal when the carbonyl group *trans* to the halogen atom vibrates.

The intensities of metal–carbon stretching bands merit more attention than they have received. There is good reason to suppose that the dipole of the metal–carbon bond is low [110], so that in the absence of any special effects, the metal–carbon stretching bands should be weak. In fact, they are moderately strong. The probable reason is once again orbital following. Since the carbon–oxygen distance changes very little during a CO stretch, the Coulomb energy of the 2π orbital remains fairly constant. What does change is the degree of overlap between metal and CO. Comparative studies of carbonyl and metal–carbonyl intensities would help to disentangle the overlap from the Coulomb energy effects. Carbonyl intensities have been reviewed [348], and the Wolkenstein bond polarisability model [351] applied to CO Raman intensities [350].

6.4. The Interpretation of Force Constants and Parameters

Interaction Constants and Parameters

As shown in Section 3.2, the interaction parameter calculated on the energy factoring model is a composite quantity, to which a range of force constants involving the MC grouping contribute. The physical interpretation of the interaction parameter presumably involves orbital following. Stretching a CO

† The interaction parameters for the species considered were not strictly comparable. Thus the unique interaction parameter i was used for substituted nickel carbonyls, and compared with the Cotton–Kraihanzel values for the *trans* parameter t in substituted molybdenum carbonyls with *trans* CO groups, and with $2c$ in species of the type *cis*-Mo $(CO)_3L_3$. Nonetheless, the results for the nickel series are unequivocal, as are those within each of the other groups considered.

group tends to cause a flow of electrons from the metal atom into the stretched CO group. As a result, π-donation from the metal to all the other ligands will tend to fall. The CO bond orders will increase, leading to a shortening of the carbon to oxygen bonds. Thus when one CO group is distorted, the others tend to change in the opposite sense, producing the positive interaction parameter observed. This parameter should be at its greatest when linking mutually *trans* CO groups, since these share two metal-donor orbitals, and experiment confirms this expectation. The true interaction constants $F_{CO,C'O'}$ probably arise by a mixture of two mechanisms. One is the orbital following mechanism, which cannot account for the fact that *cis* CO–CO interaction constants (as opposed to parameters) are greater than those between mutually *trans* groups. The other factor, which may well be more significant, is the through-space interaction of oscillating dipoles. This was suggested many years ago to explain the phenomenon of interaction [111], and has been invoked once more to account for the interactions of CO groups on different metals [112]. If two CO groups are vibrating in phase, the local oscillating dipoles set up will always repel each other, while if the vibrations are out of phase the local oscillating dipoles set up will attract each other. This effect is expected on simple electrostatic grounds to be greater for mutually *cis* carbonyls than for those that are mutually *trans* and therefore further apart. The through-space mechanism can thus account for the true interaction constants, but not for the more

TABLE 6.4. Effect of energy factoring and of neglect of anharmonicity on calculated Force Constants for $Mo(CO)_6(a, b)$.

	General Quadratic			
	Corrected (c)		Uncorrected	
	CCl_4	Vapour	CCl_4	Vapour
$F(CO)$	1715	1733	1665	1682
$F(CO, C'O'(cis))$	017	022	016	020
$F^t(CO, C'O'(trans))$	001	−006	020	012

	Energy Factored			
	Corrected		Uncorrected	
	CCl_4	Vapour	CCl_4	Vapour
$F(CO)$	1696	1715	1646	1666
$F(CO, C'O'(cis))$	029	029	027	027
$F^t(CO, C'O'(trans))$	038	029	054	045

(a) From [23].
(b) Results for $Cr(CO)_6$ and $W(CO)_6$ are similar.
(c) For anharmonicity.

complex phenomena that go to make up the energy-factored interaction parameters. Inspection of the values found for the hexacarbonyls at several levels of approximation (Table 6.4) suggests that the greater size of the anharmonic *trans* interaction parameter is due in part to anharmonicity effects, and in part to the far greater size of the *trans*, as opposed to the *cis*, MC–MC′ interaction constant. Both these effects may themselves be regarded as consequences of orbital following.†

Stretching Parameters and Force Constants

The CO stretching frequencies and force constants of carbonyl complexes are lower, sometimes markedly lower, than those of free carbon monoxide. In general, the greater the electron availability at the metal, the lower the frequency of coordinated CO. It has been customary for many years to relate the lowering of CO force constant to the degree of metal to CO π-donation, and some authors have in the past gone so far as to attempt to estimate the degree of electron back-donation [113], or to relate differences in CO parameters within molecules to specific anisotropic ligand effects [114]. There has more recently been a retreat from these extremely optimistic views, and it now seems necessary to ask how far observed spectroscopic differences do actually relate to differences in bonding, and whether the relevant changes in the bonding are in fact all of the same type.

The discussion of anharmonicity in Section 6.1 led to the conclusion that force constants or parameters calculated from anharmonic frequencies were lower by some 40–50 Nm^{-1} than those derived from the mechanical frequencies, but that the changes in this lowering from one compound to another were probable fairly small. Thus if CO stretching parameters or force constants are to be compared with similar data for species other than metal carbonyls, some allowance must be made for anharmonicity. However, if the comparison is restricted to a range of metal carbonyls, such a correction is not necessary.

The discussion of Section 3.2 shows that the CO stretching parameter gives a moderately good indication of the size of the CO stretching constant, since errors due to kinematic and dynamic coupling of the MC and CO modes tend to cancel. Thus the use of CO parameters in discussions of bonding is on the whole legitimate, though error could arise in the comparison of highly disparate species. The comparison of different kinds of CO group in

---

† The role of orbital following in increasing MC, MC′ interaction constants is evident from the general discussion given above of bonding changes during a vibration. The anharmonicity contribution is of the same order (if we accept the treatment used [94] to find the mechanical frequencies) as the harmonic interaction constant. So large a contribution could only arise by the through-space mechanism if an implausible degree of non-linearity is assumed in the bond dipoles, and in any case should not, on that mechanism, be specific to the *trans* interaction.

the same molecule presupposes the solution of a force field which is not fully determined by the all-$^{12}C^{16}O$ frequencies alone. As discussed in Chapter 3, arbitrary assumptions introduced to remove this underdeterminedness can lead to errors in the particular parameters found, though not in the average value. The usefulness of detailed *intra*molecular comparisons based on such assumptions is therefore doubtful.

A source of error that has been commonly overlooked is orbital following. The force constant associated with the distortion of any bond can, perhaps artificially, be regarded as due to two effects [115]. A certain force constant would be associated with a hypothetical bond deformation, throughout which the electron distribution (as measured, for instance, by the populations of relevant atomic orbitals) remained constant. Real processes take place with lower force constants, since changes in electron distribution take place so as to lower the energy of the distorted states. These changes in electron distribution may be identified with the orbital following effects invoked earlier to explain the high intensities and interaction parameters associated with the CO stretching modes. Interaction constants and specific CO intensities generally increase evenly along series of related compounds as CO parameters fall [108], and so it may well be that, within restricted series of compounds, orbital following effects exaggerate but do not distort the trends due to the equilibrium bonding. Comparisons between very different series are more suspect. Indeed, it has been pointed out [116] that orbital following can be treated as the response of the electronic wave function to a perturbation. The degree of orbital following therefore depends on the energies and transformation properties of the excited states. These will be different for members of different series, and indeed even for vibrations of different symmetry in the same molecule.

Two factors are now thought to be involved in that part of the CO frequency lowering that is genuinely related to bonding differences. Forward (σ) donation is expected to depopulate the antibonding 5σ orbital of CO, and thus raise the stretching frequency; such frequency raising has already been noted above for adsorbed species, and is also found in $H_3B \cdot CO$ [117] despite hyperconjugative π-donation from borane to carbonyl [118]. Calculations on such model systems as NCO^- and CO_2 [119] (compared with N_2 and N_2O) show that this σ-donation is an important phenomenon, and is responsible for the fact that N_2, which is a poorer σ-donor than CO, generally suffers a greater lowering of frequency on coordination.

The second and, despite all the above reservations, the most important factor is π-donation from the metal to the CO 2π orbitals. This effect has commonly been considered in isolation, as if it were the only factor responsible for the observed variation in ligand CO frequencies. It should be clear from this section that such reasoning is unwarranted. Its apparent success is no doubt due to the fact that the metal to carbon π-bonding and the other major factors (anharmonicity, orbital following, and reduced carbon to metal σ-donation) all depend in much the same way on the chemical constitution of the species studied. The separation of the various factors is a matter for calculation rather than experiment. Thus orbital following effects could in principle be evaluated by plotting energy against distortion, with and without re-optimising coefficients at distances different from equilibrium. Such calculations are far beyond our reach at present.

At a more empirical level, a beginning has been made with the separation of σ- and π-bonding effects by the comparison of Cotton–Kraihanzel parameters and Mulliken bond populations in octahedral metal–carbonyl halide complexes [120]. The results for sixteen different carbonyl environments fit the equation

$$k = 3642 - 1176[2\pi] - 993[5\sigma] \tag{6.8}$$

where k is the Cotton–Kraihanzel parameter for a CO group, and $[2\pi]$ and $[5\sigma]$ are the populations of the CO 2π and 5σ orbitals. The statistical fit obtained using Eq. (6.8) is highly satisfactory, although it is as yet too early to say whether the difference between the coefficients of $[5\sigma]$ and $[2\pi]$ is significant. The effects of σ- and π-bonding changes within any one complex are opposed. A CO group *trans* to halogen takes part in more metal–carbon σ-bonding ($[5\sigma]$ lower) as well as more π-bonding ($[2\pi]$ higher) than one *cis* to halogen in the same molecule. Thus the difference in force-constant underestimates the difference in π-bonding, commonly by 30% or more. On going from one complex to another, both σ- and π-bonding effects may change in the same way. Thus in $Cr(CO)_5Br^-$, Cr is at once a poorer σ-acceptor and a better π-donor than is Mn in $Mn(CO)_5Br$, and the σ-bonding changes account for between a quarter and a third of the differences in CK parameters between these two complexes. While these results may not be applicable to other species, they do at least show that changes in both σ- and π-bonding between metal and carbon are important. For a given metal, such changes can arise from variations in both σ- and π-bonding between the metal and the non-carbonyl ligands. Both of these effects are expected to be to some extent directional, affecting *trans* groups more than *cis* groups in octahedral complexes. For these reasons, any attempt to separate σ from π effects in other metal–ligand bonds from carbonyl frequency data alone appears to the author to be overambitious.

It may prove possible to separate σ and π effects in the metal–carbon bond by comparison of trends in metal–carbon bond strength with those in carbon–oxygen bonding. One would expect that, while the effects of metal–carbon σ- and π-bonding on the CO stretch are opposed, both would increase the metal–carbon force constant. Fairly rigorous force constant calculations are necessary, since individual frequencies can be misleading. For example the metal–carbon infrared band in $V(CO)_6^-$ is at markedly higher frequency than that in $Cr(CO)_6$, but the relevant force constants are very similar [27]. This is in itself a disturbing result, since the best available calculations [17] show both the metal–carbon σ- and π-bond orders to be higher in the vanadium complex.

To summarise, CO stretching frequencies, and force constants or parameters calculated from these, provide a generally reliable measure of total electron availability at the metal, although there are minor reservations when compounds of different symmetry are compared. The separation of the overall effect into its components is far more dubious, and the results of attempts to supplement CO with MC stretching data are disappointing.

6.5. The Spectra of Solids

We conclude this Chapter with a brief informal exposition of the theory used to interpret the spectra of crystalline specimens. Crystals are regular repeating arrays, which for our purposes can be regarded as of infinite extent. Crystals can be classified according to their symmetry properties as can single molecules, but there are important differences. The symmetry group of a molecule is a point group, since one particular point (the molecule's centre of mass) is invariant under all operations, and consists of proper and improper rotations only.† The symmetry group of a solid is its *space group*, and includes in addition an indefinitely large number of simple translations, as well as *screw axes* and *glide planes*. The simple translations are all built up from *primitive translations* which move the crystal through the smallest repeat unit along each of the three independent crystal axes. A parallelipiped, each of whose sides is one primitive translation, is a *primitive cell*. A primitive cell can be defined as the smallest volume repeating unit of the crystal. The choice of axes is regulated by convention, but the particular choice of primitive translations and primitive unit cell is arbitrary, and in some space groups it is possible to choose a primitive unit cell that is not a parallelipiped. A *screw axis M* is a symmetry operation that rotates the crystal through an angle $2\pi/n$ while moving it through one nth of a primitive translation. A *glide plane* is a symmetry operation that converts the crystal into its mirror image while moving it through half a primitive translation.

The space group of a crystal may be divided into two subgroups. One of these is the group of all combinations of primitive translations. The other

† Inversion and reflection may be regarded as improper rotations S_2 and S_1 respectively.

subgroup, which includes proper and improper rotations, screw axes and glide planes only, is known as the *factor group* of the crystal. Replacing the glide planes and n-fold screw axes of the factor group by reflections and n-fold rotations respectively, generates a point-group; this point group is in effect derived from the factor group by ignoring translational changes, and it is evident that it has the same multiplication table. For this reason, it is known as the *isomorphous point group*.†

The irreducible representations of the translational group are of the form

$$\Gamma\,(h, k, l) = \exp[i\mathbf{k}(h, k, l)\,.\,\mathbf{r}] \qquad (6.9)$$

where, if $\mathbf{a}$, $\mathbf{b}$, $\mathbf{c}$ are the primitive translations,

$$\mathbf{k}.\,(h\mathbf{a} + k\mathbf{b} + l\mathbf{c}) = 1. \qquad (6.10)$$

The h, k, l are positive or negative integers, and $\mathbf{k}$ is a measure of momentum. The representations of the space group and those of the factor group are related by equations of the form

$$\Gamma\,(\text{space group})\,(h, k, l; j) = \exp[i\mathbf{k}(h, k, l)\,.\,\mathbf{r}]$$
$$\times\,\Gamma(\text{factor group})\,(j). \qquad (6.11)$$

A vibrational excitation of a crystal must belong to an irreducible representation of the space group. It can be regarded as a localised excitation, belonging to a representation $\Gamma(j)$ of the factor group, propagating itself through the crystal with a momentum determined by $\mathbf{k}(h, k, l)$. For most purposes connected with vibrational spectroscopy (forward Raman scattering is an exception) excitations are only active if they have a low momentum of propagation, and the propagation factor may be ignored. Moreover, relative displacements do not affect the laws for the addition of vectors. Thus vibrations may be assigned, and selection rules derived, using the isomorphous point group alone. Tables for the factor group analysis of the vibrational spectra of solids have been published [121].

† To any operation of the factor group there corresponds an operation of the isomorphous point-group:
$$R_I = R_F \times T_R$$
Then if $P_F \times Q_F = R_F$,
$$P_I \times Q_I = (P_F \times T_P) \times (Q_F \times T_Q)$$
$$= (P_F \times Q_F) \times (T_P \times T_Q)$$
$$= (R_F) \times (\text{a translation})$$
$$= R_I.$$
Since the groups have superposable multiplication tables, they also have superposable representations and character tables.

The information to be derived from single crystal spectroscopy is in principle greater than that from disordered phases. Thus it is theoretically possible to find the direction of the electric dipole associated with an infrared-active vibration from the direction of polarisation of the light absorbed; though for reasons outlined in Section 4.4, this has not yet been accomplished for metal–carbonyl absorption bands. More usefully, if it can be established that Raman scattering by a particular mode takes place using exciting light of polarisation i, and that the Raman scattered light has the polarisation j (which may or may not be the same as i), then that mode has the same symmetry in the isomorphous point group as does the polarisability tensor component α_{ij}. The modes of crystalline solids may be correlated with those of isolated molecules through the operations of the *site group* occupied by the molecules in the crystal; this is commonly of lower order than the point group of the isolated molecule, and mixing can then take place of modes which in the isolated molecule are of different symmetry. Correlations between solution and crystal data may be used to resolve questions of assignment [122] or to decide between different possible space groups, with the usual proviso that failure to observe bands is inconclusive [123]. Energy-factored force field calculations may be carried out for the CO stretching modes in the usual way, and it is found that intermolecular interactions are comparable in size with those within a molecule. It is not yet known whether these intermolecular interactions should be thought of as due to intermolecular jostling, or to the coupling of individual molecular oscillating dipoles, and studies of the combination spectra of solids may help to resolve this point.

Chapter 7

Selected Vibrational Data for Carbonyls

7.1. Introduction

The literature on the vibrational spectra of metal carbonyls is so enormous that a selection of data might be of more use than a comprehensive catalogue. The selection presented here is meant to serve three purposes. One is to illustrate and exemplify the general points raised in the previous chapters. The second is to enable users of this book to place spectroscopic data for particular chemical species in context by comparison with data for a range of related species. The third aim is to provide the reader with information with which to test his views on the chemical significance of vibrational data, and hence to suggest new correlations and comparison. The choice of papers cited is necessarily arbitrary, and much excellent work, particularly early work, has had to pass without mention.

7.2. Simple Model Systems

Carbon Monoxide

The fundamental, binary, ternary and quarternary bands of carbon monoxide ($^{12}C^{16}O$) gas have been analysed, and much information is also available for ^{13}CO. The observed fundamental frequencies and calculated harmonic frequencies [124, 125, 126] are, for ^{12}CO,

$$v = 2143 \, \text{cm}^{-1} \quad \omega = 2170 \, \text{cm}^{-1}$$

and for ^{13}CO,

$$v = 2096 \, \text{cm}^{-1} \quad \omega = 2121 \, \text{cm}^{-1}.$$

The ratio of the observed frequencies is 1:0·9780, while the harmonic frequencies are in the ratio 1:0·9776. The ratio predicted from the effective masses is 1:0·9777. Thus the observed frequencies are slightly closer together than the mechanical frequencies. This is as expected; the band due to $^{12}C^{16}O$ has a higher amplitude than that due to $^{13}C^{16}O$, and is therefore liable to be more affected by anharmonicity.

Borane Carbonyls

The carbonyl stretching frequency of $H_3B.CO$ is $2165 \, cm^{-1}$ [117]. This is higher than for carbon monoxide itself, thus presumably showing the effects of σ-bonding from carbon. The corresponding frequency of tetraborane carbonyl, B_4H_8CO, is $2150 \, cm^{-1}$ [127].

Triatomic Molecules XCO

The vibrational frequencies of a variety of linear species in which CO is bonded to one other atom are presented in Table 7.1. In CO_2 and NCO^-, the two stretching modes both involve roughly equal distortions of the two bonds to carbon, and the frequencies are not directly comparable with those of CO or of metal carbonyls. In SCO, v_2 is mainly a CO stretch, but is raised in frequency by kinematic coupling with the SC stretch. In HCO, the insensitivity of the higher frequency to isotopic substitution at carbon shows it to be almost a pure H—C stretch. The lower frequency then corresponds fairly closely to a pure CO stretch. The HCO radical contains one electron in excess of those required to fill the bonding orbitals, and this presumably is in an orbital closely related to 2π of free CO. Thus the observed frequency corresponds to a CO valence bond order of 2·5, with the carbon being σ-bonded to another atom.

TABLE 7.1. Vibrational frequencies (cm^{-1}) of some linear triatomic species XCO (a).

Species (b)	v_1	v_2 (c)	v_3
$H^{12}CO$	2488 (d)	1090	1861
$H^{13}CO$	2488	1084	1821
NCO^- (e)	1207 (f)	637,628	2165
CO_2	1337 (f)	677	2349
SCO	859	524	2064

(a) Data from [126.]
(b) As vapour.
(c) Bending Mode.
(d) v (C—H).
(e) As solid potassium salt.
(f) Symmetric stretch.

Adsorbed CO

Some frequencies found for CO adsorbed onto a range of metal and metal oxide supports are presented in Table 7.2. The CO frequency depends on the nature of the support and the pre-treatment of the surface. Thus the frequency of CO adsorbed onto zinc oxide is high, presumably because the substrate cannot act as a π-donor. The frequency of CO adsorbed on copper is increased by surface oxidation, but even when the metal has been exhaustively reduced, it is higher than the corresponding frequency found over nickel, palladium or platinum supports. In the cases of nickel and palladium, two groups of frequencies are found, separated by 125 cm^{-1} or more, and the lower frequency group is attributed to adsorbed CO molecules that bridge two or more substrate metal atoms.

Table 7.2. Vibrational Frequencies (cm^{-1}) for adsorbed CO

Substrate	Frequencies	Reference
Ni	1908, 2033	[128]
Pd	1916, 2053	
Pt	2070	
Cu	2128	
ZnO	2212, 2200, 2187	[129]
Pt-SiO$_2$	2100, 2092, 2085	
Cu$-$SiO$_2$	2143 to 2130 (a)	
Cu$-$SiO$_2$ oxidised (b)	2136	

(a) Observed frequency lowered by successive hydrogen reduction–outgassing cycles.
(b) Uptake of oxygen one atom per atom of Cu.

7.3. Mononuclear Metal Carbonyls

Investigations of the spectra of the simple metal carbonyls stretch back over more than twenty years. In this Section and its accompanying Tables reference will be made to some of the more recent studies; the many excellent earlier studies can usually be traced through these.

V(CO)$_6$

Studies of hexacarbonyl vanadium have been inhibited by its high air sensitivity, and Raman studies in particular have been impeded by its colour. Vanadium hexacarbonyl is a 17-electron (t_{2g}^5) system, and is in principle expected to show a Jahn–Teller distortion that lowers its symmetry from that of a regular octahedron. Such a distortion has not been detected spectroscopically, and the deviations from regularity may well be very small. It is, however, probably relevant that the vapour phase T_{1u} $v(CO)$ band of $V(CO)_6$ is broad compared with that of $Cr(CO)_6$, and that P, Q and R structures cannot be resolved [130].

$V(CO)_6^-$

The Raman and infrared spectra of the $[V(CO)_6]^-$ anion have been investigated [27] in *sym*-tetrachloroethane, acetonitrile, and the solid state, using $(CH_3)_4N^+$ and $Na(diglyme)_2^+$ as cations. Frequencies v_1-v_{12} were assigned from their positions, infrared and Raman activities in solution and occurrence in combination bands. The harmonic frequencies ω_1, ω_3, ω_6 were estimated by comparing the anharmonic defects of the combination bands $(v_1 + v_6)$, $(v_3 + v_6)$ with anharmonicity data for $Cr(CO)_6$ (below). This is not a satisfactory procedure, but seems better than applying no correction at all. v_{13}, a low frequency totally inactive (T_{1u}) bending mode, could not be located. The frequencies found are given in Tables 7.3 and 7.4; it is worth mentioning that v_1 and v_3 could readily be observed in the infrared spectrum, and that Raman scattering by v_1 was remarkably weak.

TABLE 7.3. CO stretching frequencies (a) (cm^{-1}) of the hexacorbonyls

Species	Solvent	$v_1(A_{1g})$	$v_3(E_g)$	$v_6(T_{1g})$	Reference
$V(CO)_6$	Vapour			1987	[130]
	C_6H_{12}			1976	[131]
$V(CO)_6^-$	$C_2H_2Cl_4$	2020	1895	1858	[27]
	CH_3CN	2020	1894	1858	[27]
	solids	1995	1870		[27]
	CH_2Cl_2			1845	[27]
	acetone			1860	[131]
	THF (b)			1843	[132]
$Cr(CO)_6$	Vapour	2118·7	2026·7	2000·4	[23]
	Vapour, ω (c)	2139·2	2045·2	2043·7	[23]
	CCl_4	2112·4	2018·4	1984·4	[23]
	CCl_4, ω (c)	2132·9	2036·9	2027·2	[23]
	Solid	2109·9	2006·1		[23]
	Solid, ω (c)	2130·4	2024·6		[23]
	Ar (d)			1995·2, 1990·0 (e)	[44]
	Hc (f)	2113 (g)	2018 (g)	1987	[134]
	CH_2Cl_2	2110	2018		[133]
$Mo(CO)_6$	Vapour	2120·7	2024·8	2003·0	[23]
	Vapour, ω (c)	2144·2	2043·3	2043·1	[23]
	CCl_4	2116·7	2018·8	1986·1	[23]

TABLE 7.3—*contd.*

Species	Solvent	$v_1(A_{1g})$	$v_3(E_g)$	$v_6(T_{1u})$	Reference
Mo(CO)$_6$	CCl$_4$, ω (c)	2140·2	2037·3	2026·2	[23]
	Solid	2113·6	2005·2		[23]
	Solid, ω (c)	2137·1	2023·7		[23]
	Ar (d)			1997·3, 1992·2 (e)	[44]
	Hc (f)	2116 (g)	2019 (g)	1989	[134]
		2116·5(g)	2018·9(g)	1989·3	[29]
	CH$_2$Cl$_2$	2117	2019		[133]
W(CO)$_6$	Vapour	2126·2	2021·1	1997·6	[23]
	Vapour, ω (c)	2153·2	2037·6	2037·6	[23]
	CS$_2$	2116·6	2009·8	1976·6	[23]
	CS$_2$, ω (c)	2143·6	2026·3	2016·6	[23]
	Solid	2115·3	1998·4		[23]
	Solid, ω (c)	2142·3	2014·9		[23]
	Ar (d)			1992·0, 1986·6 (e)	[44]
	Hc (f)	2120 (g)	2013 (g)	1986	[134]
	CH$_2$Cl$_2$	2120	2012		[133]
Mn(CO)$_6$+	THF (b)			2094	[132]
Re(CO)$_6$+	CH$_3$CN	2197	2122	2085	[27]
	CH$_3$CN, ω (h)	2224	2139	2131	[27]

(a) Observed (anharmonic) frequencies unless otherwise stated.
(b) Tetrahydrofuran.
(c) Estimated harmonic frequency.
(d) Solid matrix.
(e) Lower frequency component of unresolved doublet.
(f) Hydrocarbon.
(g) Inferred from spectrum of $M(^{12}CO)_5{}^{13}CO$.
(h) Anharmonic corrections estimated from the values for Cr(CO)$_6$, W(CO)$_6$.

Hexacarbonyls of Group VI; Cr(CO)$_6$, Mo(CO)$_6$, and W(CO)$_6$

Infrared spectra, including binary and ternary CO solution spectra, of all three of the Group VI hexacarbonyls have been reported both in the vapour phase and in solution (CCl$_4$ for Cr(CO)$_6$ and Mo(CO)$_6$; CS$_2$ for W(CO)$_6$). Raman data have also been obtained for the solids, and for solutions in the same solvents as those used for the infrared study. Similar data were also collected for the species $M(C^{18}O)_6$ and $M(^{13}C^{16}O)_6$, and the results used to calculate "rigorous" force constants [23]. The elusive frequency v_{13} was found from analysis of weak combination bands. Coriolis coupling constants were estimated from the separation of P and R branches in the vapour phase spectra. The anharmonic corrections to the carbonyl stretching modes were derived using the binary and ternary combination spectra. The frequency

TABLE 7.4. Fundamental frequencies of hexacarbonyls [23, 27].

Assignment	$V(CO)_6^-$	$Cr(CO)_6$		$Mo(CO)_6$		$W(CO)_6$		$Re(CO)_6^+$
	CH_3CN	CCl_4	Vapour	CCl_4	Vapour	CS_2	Vapour	CH_3CN
A_{1g} ν_1(CO)	2020	2112·4	2118·7	2116·7	2120·7	2118·6	2126·2	2197
ν_2(MC)	374	381·2	379·2	402·2	391·2	427·1	426	441
E_g ν_3(CO)	1894	2018·4	2026·7	2018·8	2024·8	2009·8	2021·1	2122
ν_4(MC)	393	394	390·6	392	381	412	410	426
T_{1g} ν_5(δMCO)	356		364·1		341·6		361·6	354
T_{1u} ν_6(CO)	1858	1984·4	2000·4	1986·1	2003·0	1976·6	1997·6	2085
ν_7(δMCO)	650	664·6	668·1	592·8	595·6	583·1	586·6	584
ν_8(MC)	460	443·8	440·5	367·0	367·2	374·4	374·4	356
ν_9(δCMC)	92 (b)	103	97·2	91	81·6	92·0	82·0	82
T_{2g} ν_{10}(δMCO)	517		532·1		477·4		482·0	486
ν_{11}(δCMC)	84	100·8	89·7	91	79·2	92·0	81·4	82
T_{2u} ν_{12}(δMCO)	506		510·9		507·2		521·3	522
ν_{13}(δCMC)			67·9		60		61·4	
ω_1 (a)		2132·9	2139·2	2140·2	2144·2	2143·6	2053·2	
ω_3		2036·9	2045·2	2037·3	2043·3	2026·3	2037·3	
ω_6		2027·2	2043·7	2026·2	2043·1	2016·6	2037·6	

(a) Estimated harmonic frequencies; see text.
(b) $C_2H_2Cl_4$

$3\nu_6$ is expected to be split into components by an amount proportionate to the parameter G_{66} of Eq. (6.6). This splitting was observed in carbon disulphide solutions of $W(CO)_6$, and led to a value of 4·0 cm for G_{66}. No splitting could be observed in the corresponding bands of $Cr(CO)_6$ and $Mo(CO)_6$ in carbon tetrachloride solution. Nevertheless, a value of $5\,cm^{-1}$ was assumed for G_{66} in these latter carbonyls, since the other anharmonicity constants are higher than in $W(CO)_6$.

The frequencies found for the $^{12}C^{16}O$ derivatives are given in Tables 7.3 and 7.4; frequencies for the $^{13}C^{16}O$ and $^{12}C^{18}O$ derivatives are given in the original paper. The assignment of the frequencies is fairly straightforward. The carbonyl stretching modes ν_1, ν_3 and ν_6 are well separated from the others, and may be assigned from their respective Raman and infrared activities, from the depolarisation of Raman scattering by ν_3 while ν_1 is polarised, and from their relative frequencies. The metal-carbon stretching mode ν_2 is the only other polarised Raman band, and ν_4 (Raman-active, depolarised) and ν_8 (infrared-active) are readily distinguished from MCO bending modes by their behaviour on isotopic substitution. The infrared-active bending modes ν_7 and ν_9, and the Raman-active modes ν_{10} and ν_{11}, may be divided into MCO and CMC bends on the basis of their frequencies. In some cases, frequencies for one phase have been estimated from combination bands, using fundamental bands in other phases to specify the assignments. The MCO frequencies ν_5 and ν_{12} can only be observed in combination, as can ν_{13}. ν_5 and ν_{12} occur in combination with different modes and are thus distinguished, while ν_{13} is distinguished from ν_{12} by its frequency. Minor anomalies arose in the derivation of force constants for $Cr(CO)_6$ from the data, and the calculated Coriolis coupling constant ζ_9 did not agree with experiment. It was concluded that the observed frequencies for $Cr(CO)_6$ are distorted, probably by Fermi resonance between ν_8 and $(\nu_5 + \nu_9)$, and the best force constant analysis for $Cr(CO)_6$ makes use of ζ_9 for $Cr(CO)_6$ and of force constants found for the other hexacarbonyls. The results [23] may be taken as definitive, subject only to the reservations expressed in Chapter 6 about the procedure used [94] to determine the anharmonicity corrections.

The infrared spectra of the isotopically labelled species $M(^{12}CO)_5^{13}CO$ have been examined in the carbonyl region in hydrocarbon solution, using either natural abundance [134] or ^{13}CO enrichment [29]. As explained in Section 3.6, the labelled species have five frequencies. Two of these (E and a mode designated by both groups of authors as B_1, although B_2 is equally correct) correlate directly with the modes T_{1u} and E_g of the hexacarbonyls. The remaining three bands, of symmetry A_1 in C_{4v}, have been observed in all cases. There is also a weak, broad absorption around $2080\,cm^{-1}$ that represents one or more combination bands. The frequencies found are given in Table 7.5, and the frequencies inferred for the parent hexacarbonyls are included in Table 7.3.

TABLE 7.5. Spectra (a) (cm^{-1}) of Species M (^{12}CO)$_5$^{13}CO.

Species	Phase	$A_1(1)$ (b)	$\delta(1)$(c)	$A_1(2)$	$\delta(2)$	A_1 (3)	$\delta(3)$	References
Cr(^{16}CO)$_5$^{13}CO	Vapour					1967·2	33·2	[23]
	H(d)	2107	6	2010	8	1956	31	[134]
	Ar					1958·5	31·5 (e)	[44]
Mo(^{12}CO)$_5$^{13}CO	Vapour					1970·1	32·9	[23]
	H	2110	6	2011	8	1957	31	[134]
		2110·6	5·9	2010·9	8·0	1958·0	31·3	[29]
	Ar					1960·7	31·5	[44]
W(^{12}CO)$_5$^{13}CO	Vapour					1964·9	32·7	[23]
	H	2114	6	2004	9	1954	32	[134]
	Ar					1955·8	30·8	[44]

(a) For B_1 (B_2) and E frequencies, see E_g and T_u of Table 7.3.
(b) $A_1(1)$, $A_1(2)$ and $A_1(3)$ are the observable bands, in order of decreasing frequency.
(c) "Isotope shifts" relative to M(^{12}C^{16}O)$_6$;$\delta(1) = \nu(A_{1g}) - \nu(A_1(1))$.
(d) Hydrocarbon solution.
(e) Relative to lower component of split T_{1u} band.

[Mn(CO)$_6$]$^+$ and [Re(CO)$_6$]$^+$

The infrared spectrum of Mn(CO)$_6$$^+$ in the carbonyl region has been reported [132]. As expected, a single strong band is observed at relatively high frequency (Table 7.3). The infrared and Raman spectra of Re(CO)$_6$$^+$ have been studied in more detail [27], and the results are compared with those for W(CO)$_6$ and related species in Tables (7.3, 7.4). The anharmonic corrections for the binary combination carbonyl spectra of Re(CO)$_6$$^+$ are very similar to those of Cr(CO)$_6$ and W(CO)$_6$, and the anharmonicity corrections for Re(CO)$_6$$^+$ were assumed to be the same as those for the neutral hexacarbonyls. A force constant analysis of Re(CO)$_6$$^+$ has been carried out [135] using the data of [27].

Pentacarbonyls of Group VI

The pentacarbonyls of Group VI have been generated photochemically in argon matrices [44]; the spectra (Table 7.6) indicate C_{4v} symmetry, the highest A_1 frequency being extremely weak. Similar species are generated in hydrocarbon matrices [43, 45], but the infrared frequencies are somewhat depressed relative to the argon matrix results, and an electronic absorption band has moved to higher frequency. There is thus a strong interaction between the pentacarbonyl fragment and the hydrocarbon solvent, and some authors [81] regard this interaction as constituting a chemical bond.

Included in Table 7.6 are data for a species formed on softening of a hydrocarbon lattice containing $Mo(CO)_5$ (C_{4v}). This species was originally formulated as a trigonal bipyramidal isomer of $Mo(CO)_5$ [43], but is has since been argued that it is in fact polymeric [45].

TABLE 7.6. CO frequencies for photochemically generated group VI pentacarbonyls

Species	Matrix	Frequencies	Reference
$Cr(CO)_5$	H	$2088(A_1), 1955(E), 1928(A_1)$	[43]
	Ar	$2093 \cdot 4, 1965 \cdot 4, 1936 \cdot 1 \ (1933 \cdot 5w \ ^{13}CO)$	[44]
$Mo(CO)_5$	H	$2093(A_1), 1960(E) \ 1920(A_1)$	[43]
	Ar	$2098 \cdot 0, 1972 \cdot 7, 1932 \cdot 6 \ (1929 \cdot 5w \ ^{13}CO)$	[44]
$Mo(CO)_5 D_{3h} (?)$	H (a)	2007, 1995	[43]
$W(CO)_5$	H	$2092(A_1), 1952(E), 1924(A_1)$	[43]
	Ar	$2097 \cdot 3, 1963 \cdot 3, 1932 \cdot 2 \ (1928 \cdot 9w \ ^{13}CO)$	[44]

(a) Softened matrix. See text for criticism of assignment.

$Mn(CO)_5^-$

The anion $Mn(CO)_5^-$ is isoelectronic with $Fe(CO)_5$ (discussed below) and is presumed to have the same trigonal bipyramidal structure. The spectroscopic data available are consistent with this assumption, and are summarised in Table 7.7.

Iron Pentacarbonyl

Despite some earlier controversy the structure of pentacarbonyl iron has been unequivocally established as trigonal bipyramidal [136]. There are thus two different kinds of CO group, axial and equatorial. To a first approximation, these two kinds of group may be considered as vibrating separately, but when the vibrations of the two kinds of CO group span the same irreducible representation of the point group D_{3h}, coupling can occur and may be significant.

CO Stretching Modes. The CO stretching modes obtained in a variety of solvents are shown in Table 7.7, together with data for isotopically substituted species. The energy factored force field for $Fe(CO)_5$ is underdetermined, even using the Raman-active frequencies. The force field has, however, been solved by Bor [137], using data for $Fe(CO)_5$ enriched in $Fe(^{12}CO)_4 {}^{13}CO$. There are two possible isotopically labelled species, since

TABLE 7.7. CO stretching frequencies for some pentacarbonyls.

Species	Solvent	$v_1(A_1')$	$v_2(A_1')$	$v_6(A_2'')$	$v_{10}(E')$	Ref.	Comments
Fe(CO)$_5$	(solid)	2115	2033	2003	1982	[138]	IR
		2117	2022		1999, 1971	[139]	R
	(liquid)			2002	1979	[139]	IR
		2116	2030		1989	[139]	R
	(vapour)			2034	2012	[138]	IR
	C$_6$H$_{12}$			2022	2000	[139]	IR
	CCl$_4$			2020	1995	[138]	IR
Fe(^{12}CO)$_4$^{13}CO(ax)	C$_7$H$_{16}$	2105·5 (A_1), 1987·4(A_1) (a)				[137]	IR
Fe(^{12}CO)$_4$^{13}CO(eq)	C$_7$H$_{16}$	2108·0 (A_1), 1964·0(A_1) (a)				[137]	IR
Ru(CO)$_5$	C$_7$H$_{16}$			2035	1999	[140]	IR
Os(CO)$_5$	C$_7$H$_{16}$			2034	1991	[140]	IR
	(vapour)			2047	2006	[140]	IR
Mn(CO)$_5^-$	THF, diglyme			1898	1863	[141]	IR

(a) Other bands obscured by Fe (^{12}CO)$_5$ spectrum.

the labelled ligand may occupy either an axial or an equatorial site. Of the large number of possible bands for these species, only four were observed. These, together with the two infrared-active frequencies of Fe(^{12}CO)$_5$, suffice to determine six parameters. The calculation performed was equivalent to the determination of the effective ratio of reduced masses of ^{12}C^{16}O and ^{13}C^{16}O, and of the five parameters of the energy factored force field. The ratio of reduced masses is calculated in the energy factoring approximation to be 0·9776, but (as shown in Section 3.1) the observed frequencies are better fitted by an "empirical" ratio (of 0·9772). The calculated parameters are, in the notation of Section 3.6.,

$$k(e) = 1657 \, \text{Nm}^{-1}$$

$$k(a) = 1695 \, \text{Nm}^{-1}$$

$$e = 28 \quad \text{Nm}^{-1}$$

$$f = 40 \quad \text{Nm}^{-1}$$

$$t = 41 \quad \text{Nm}^{-1}$$

It is noteworthy that the equatorial CO groups have a lower force constant than does the axial group. It is also noteworthy that the predicted stretching parameters for a pure axial A_1' mode and a pure equatorial A_1' mode are $1736\,\mathrm{Nm}^{-1}$ and $1737\,\mathrm{Nm}^{-1}$ respectively. In other words, the actual A_1' modes are almost exactly in-phase and out-or-phase combinations of the axial and equatorial A_1' modes, and cannot be assigned exclusively to either set of CO groups. The separation of the actual A_1' modes is due entirely to the interaction constant connecting them, which in this case takes the largest value consistent with the $Fe(^{12}C^{16}O)_5$ frequencies. This is unlikely to be a general result, however, even for carbonyl complexes of neutral iron. The stretching parameters for the pure axial and equatorial A_1' modes in $Fe(CO)_5$ take the forms $k(a) + t$ and $k(e) + 2f$. In the axially monosubstituted derivatives, the equatorial stretching parameter is unaltered, but the axial parameter reduces to $k(a)$. Moreover, $k(a)$ itself is expected to be more sensitive to substitution than is $k(e)$. It follows that, while $Fe(CO)_5$ shows maximum mixing of the A_1' modes and no separability of axial and equatorial motions, in $Fe(CO)_4L$ there is less than maximum mixing, and the lower frequency A_1 band corresponds principally to the vibration of the axial carbonyl.

Other bands. The remaining bands of $Fe(CO)_5$ are assignable by the use of correlations with species $Fe(CO)_4L$ and $Fe(CO)_3L_2$, combination and overtone bands, and Raman scattering depolarisation data. Photolysis by the Raman-exciting source no longer presents a problem, thanks to the availability of laser instruments with low frequency output. The first harmonic of the helium-neon laser occurs at $15{,}780\,\mathrm{cm}^{-1}$, and has been used successfully on liquid and solid samples of $Fe(CO)_5$. The species L chosen for this work [139] were trimethylphosphine, and its arsine and stibine analogues, because the internal vibrations of these ligands, and the vibrations of the metal–ligand skeletons, are in regions that do not interfere with the spectra of the metal carbonyl skeletons.

In the middle frequency region, two polarised bands are observed in the Raman spectrum. These must be the two fully symmetric metal-carbon stretching bands. One of these correlates with strong bands of similar frequency in derivatives $Fe(CO)_4L$ and $Fe(CO)_3L_2$ and is therefore assigned as the equatorial A_1' metal-carbon stretching band. The other is far weaker, and correlates with a band of far higher frequency in $Fe(CO)_4L$, and absent in $Fe(CO)_3L_2$. Accordingly, this band is assigned to the axial A_1' metal-carbon stretch; its increase in frequency on going from $Fe(CO)_5$ to $Fe(CO)_4L$ is too great to be assigned to kinematic effects alone, and presumably reflects the trans-weakening influence of CO. Of the remaining frequencies between 550 and $300\,\mathrm{cm}^{-1}$, that at $432\,\mathrm{cm}^{-1}$ is infrared-active but not Raman-active, and must therefore be assigned to the asymmetric (A_2'') axial metal-carbon stretch, while the strong infrared band at $475\,\mathrm{cm}^{-1}$ also appears in the

Raman spectrum, and is therefore assigned to the asymmetric (E') equatorial metal-carbon stretching motion. Correlations with $Fe(CO)_4L$ and $Fe(CO)_3L_2$ confirm the latter assignment. It should be noted that the infrared-Raman comparison used to derive these assignments depends on the improved resolution made possible by Raman spectrometers, and on the observation of splittings at low frequencies in apparently simple bands, and differs in detail from earlier interpretations [142, 143].

The assignments of the FeCO bending modes are based on similar arguments. Thus the equatorial A_2'' mode is only infrared-active, the equatorial and axial E'' modes are only Raman-active, and the equatorial and axial E' modes show both kinds of activity. Axial and equatorial modes of the same symmetry are distinguished, to the extent that such a distinction is meaningful, by correlation with the spectrum of $Fe(CO)_4L$. The remaining bending mode, A_2' (equatorial), is totally inactive, but is thought by analogy with related bands in $Ni(CO)_4$ to be responsible for a strongly polarised Raman overtone

TABLE 7.8. Fundamental frequencies of $Fe(CO)_5$ [139].

Mode	IR frequency (cm^{-1})			Raman frequency (cm^{-1})		
	Liquid	C_6H_{12}	THF	Liquid	Solid	THF
$A_1': v_1[CO(ax) + CO(eq)]$				2116	2117	
$v_1[CO(ax) - CO(eq)]$				2030	2022	
$v_3[MC(eq)]$				418	425	418
$v_4[MC(ax)]$				381	380	377
$A_2': v_5[\delta MCO(eq)]$ (a)				278	279	279
$A_2'': v_6[CO(eq)]$	2002	2022				
$v_7[\delta MCO(eq)]$	615	615				
$v_8[MC(ax)]$	432	430	431			
$v_9[\delta CMC(eq)]$	72					
$E': v_{10}[CO(eq)]$	1979	2000		1989	$\left\{\begin{matrix}1999\\1971\end{matrix}\right.$	
$v_{11}[\delta MCO(eq)]$	637	642	653	653	658	653
$v_{12}[\delta MCO(ax)]$	554	553	553	559	562	560
$v_{13}[MC(eq)]$	475	475	475	482	490	478
$v_{14}[CMC(eq)]$	112	114		107	$\left\{\begin{matrix}118\\113\end{matrix}\right\}$	107
$v_{15}[CMC(ax)]$				64	$\left\{\begin{matrix}73\\57\end{matrix}\right\}$	64
$E'': v_{16}[MCO(eq)]$				491	496	488
$v_{17}[MCO(ax)]$				448	450	448
$v_{18}[CMC(ax)]$					132	

(a) $[2v_5]/2$.

band at $758 \, \text{cm}^{-1}$. The weakness in the infrared spectrum of the E'(ax) deformation can be explained by coupling to E'(eq), which is correspondingly enhanced.

The observation and assignment of the CFeC bending modes has presented some difficulty. The vibrations are of low frequency and show splitting and marked temperature shifts in the solid. In addition, some of the infrared-allowed bands are actually extremely weak. Nonetheless, the method of correlation and the use of infrared-Raman coincidences lead unequivocally to the assignments of Table 7.8. The only possible ambiguity is in the relative assignments of E'(ax) and E''(ax) in $Fe(CO)_5$, and the assignment given is based on analogy with the frequency ordering for the metal-phosphorus deformations in $Fe(CO)_3(PMe_3)_2$ [139].

Pentacarbonyls of Ruthenium and Osmium

$Ru(CO)_5$ and $Os(CO)_5$ are presumably isostructural with $Fe(CO)_5$. Infrared spectra have been obtained, and are assignable by analogy with $Fe(CO)_5$. The values found are included in Table 7.8.

Tetracarbonyl Nickel

The frequency assignments for $Ni(CO)_4$ are for the most part straightforward, and are given in Table 7.9. The fully symmetric vibrations are infrared-inactive, and are Raman active and strongly polarised. The NiCO bending mode of symmetry E has not been observed directly, but has been located by extrapolation from bands in the spectra of substituted species [144]. The CNiC mode of symmetry E has been observed in the Raman spectrum but is infrared inactive. Four modes are infrared active, and of these the asymmetric CO stretching mode and the CMC bending mode of T_2 symmetry are immediately assignable from their frequencies. There remain bands at 454 and $417 \, \text{cm}^{-1}$, of which the former is assignable as an NiCO deformation while the latter is due to the asymmetric NiC stretching mode. This assignment follows from the positions of the corresponding bands of isotopically labelled species (compare Section 3.2). Finally, the NiCO band of symmetry T_1 is totally inactive; it is assigned from its binary overtone band which occurs weakly, with a high degree of polarisation, in the Raman spectrum (compare $2A_2'$ of $Fe(CO)_5$).

Attempts to fit data for $Ni(CO)_4$ and for its fully isotopically labelled analogues to a generalised quadratic valence force field lead to difficulties, since the frequencies observed do not obey the Teller–Redlich product rule. It has been suggested that the failure to obey the rule is due to distortions of the band envelopes by thermally excited species [99], or to differences in force field between isotopically different molecules [100]. This latter proposal is rather surprising, but the suggested discrepancy could arise as a result of different vibrational amplitudes and anharmonicity corrections.

TABLE. 7.9. Fundamental frequencies (cm^{-1}) of some species $M(CO)_4$

Species	$[Fe(CO)_4]^{2-}$		$[Co(CO)_4]^{-}$		$Ni(CO)_4$			$Pd(CO)_4$	$Pt(CO)_4$
solvent	H_2O	DMF(b)	Dry THF	wet THF(c)	vapour	CCl_4	CO matrix	CO matrix	CO matrix
reference	[145]	[146]	[355]	[355] [99]; [100] agrees closely			[343]	[343]	[343]
A_1: ν_1(CO)	1788?(a)	2002			2132·4	2125·0	2130	2122	2119
ν_2(MC)		431			370·8	379·8			
E: ν_3(δMCO)					380	380			
ν_4(δCMC)					62	78			
T_2: ν_5(CO)	1786	1888	1889(s,sh), 1886(vs), 1858(m)(a)	1886	2057·8	2044·5	2043	2066	2049
ν_6(δMCO)	556	523			458·9	455			
ν_7(MC)	646	556			423·1	422·5			
ν_8(δCMC)					80	91			
T_1: ν_9(δMCO)						300			
ω_1					2154·1	2146·7			
ω_5					2092·2	2079·0			

(a) See text
(b) DMF = N, N-dimethylformamide
(c) 8% v/v H_2O

Whatever the explanation, the failure to obey the product rule leads to difficulties in calculating the force field. Calculations designed to find a field that fits the frequencies for $Ni(^{12}C^{16}O)_4$, $Ni(^{13}C^{16}O)_4$, and $Ni(^{12}C^{18}O)_4$ fail to converge on a solution, while the analytic determination of symmetry force constants, using the rules for isotopic shifts as a constraint, gives results that are clearly unsatisfactory. Acceptable force fields can, however, be obtained by making use of band contours to determine Coriolis coupling constants, while constraining certain unimportant interaction constants to be zero. An alternative method uses an extension of the technique of correlation of series of compounds. Symmetry force constants (i.e. effective force constants for symmetry coordinates, containing bond force constants and bond interaction constants) have been found for substituted nickel carbonyls, ignoring the symmetry interaction constant that links symmetry correct combinations of M—C and C—O stretching modes. Extrapolation of values for species $Ni(CO)_2L_2$ and $Ni(CO)_3L$ leads to a value for the T_2 symmetry force constant for metal-carbon stretching. This force constant together with frequency data for $Ni(CO)_4$ is used to calculate a value of the symmetry interaction constant. The symmetry force constants of the substituted derivatives can then in turn be recalculated, using the value found earlier for the interaction force constant, and extrapolated to give an improved extimate of the symmetry force constant in $Ni(CO)_4$ itself. A knowledge of the symmetry force constants is enough to define the values of the force constants in terms of individual bond displacements. The force fields found by these two very different methods are in remarkably good agreement [99, 100].

Palladium and Platinum Tetracarbonyls

The elusive tetracarbonyls of palladium and platinum have recently been characterised as products of the co-condensation of metal vapours and CO at a refrigerated window, and the infrared and Raman spectra compared with those for tetracarbonyl nickel in the some matrix; frequencies are included in Table 7.9. In all cases, the highest observed CO stretching bond is Raman-active only and polarised, while the other observed frequency occurs in both sets of spectra, being Raman-depolarised. This is as expected for the A_1 and T_2 modes of the anticipated tetrahedral structures. The data show the usual alternation on descending the group, the tetracarbonyl of platinum being intermediate between those of palladium and nickel [343].

Tetracarbonylferrate (-2) and Tetracarbonylcobaltate (-1), $Fe(CO)_4{}^{2-}$ and $Co(CO)_4{}^-$

Data for these species [140, 145, 146] are included in Table 7.9. The pronounced shift of the CO stretching bands to low frequency, in

TABLE 7.10. CO stretching frequencies of $Mn_2(CO)_{10}$ and Related Species.

Species	Solvent (a)	Method (b)	$\nu_1(A_1)$	$\nu_2(A_1)$	$\nu_{11}(B_1)$	$\nu_{12}(B_1)$	$\nu_{17}(E_1)$	$\nu_{25}(E_2)$	$\nu_{31}(E_3)$	Reference	Comments
$Mn_2(CO)_{10}$	H	R	2116	1997	2045·8			2024	1981	[122]	(c)
	H	IR;^{13}C	*2115·0*	*1997·5*	1983·8		*2014·7*	*2023·0*	*1981·5*	[147]	(d, e)
$Re_2(CO)_{10}$	H	R	2128	1993				2029		[122]	(c)
	H	IR			2070	1976	2014			[148]	
$MnRe(CO)_{10}$	H	IR	2124		2054	1978	2017	2031		[148]	(c, f)
$Tc_2(CO)_{10}$	H	IR			2065	1984	2017			[148]	
$Cr_2(CO)_{10}^{2-}$	MeCN	IR	1983*vw*		1917	1793·5*sh*	1890			[149]	(c)
$Mo_2(CO)_{10}^{2-}$	MeCN	IR	1983*vw*		1937	1795	1895			[149]	(c)
$W_2(CO)_{10}^{2-}$	MeCN	IR	1978*vw*		1943	1794·5	1893·5			[149]	(c)

(a) H = saturated hydrocarbon.
(b) R = Raman, IR = infrared, ^{13}C = use of data for ^{13}CO-enriched species to calculate positions of IR-inactive frequencies.
(c) Medium frequency data also presented and discussed.
(d) Inferred frequencies in italics.
(e) for criticism of assumptions used, see text.
(f) Assignment made [375] with hindsight. Additional frequencies at 2044 cm^{-1} (*vw*) and 1945 cm^{-1} (*w*) appear to be ^{13}CO satellites of B_2 modes.

comparison with $Ni(CO)_4$, is as expected. More puzzling is the failure to resolve the A_1 and T_2 carbonyl bands of $Fe(CO)_4^{2-}$ in the Raman spectrum. It may be that the A_1 band was too weak to be observed with the non-laser source used; if this is the correct explanation, the situation is closely analogous to that in $V(CO)_6^-$ [27], and the ion would well repay examination using laser instrumentation.

The A_1 and T_2 $v(M-C)$ modes of $Co(CO)_4^-$ both occcur at higher frequency than the corresponding bands of $Ni(CO)_4$, so that the $v(M-C)$ T_2 mode, which may be recognised by its high infrared intensity, actually occurs at higher frequency than the MCO bending mode of the same symmetry. It appears that in this case the increase in π-back-bonding on going from $Ni(0)$ to $Co(-1)$ more than compensates for the decrease in carbon to metal σ-donation.

In solvents of low polarity, $NaCo(CO)_4$ exists as tight ion pairs of symmetry C_{3v}, the higher A_1 mode becomes infrared allowed, and the T_2 CO stretching mode is split into two components. Tight ion pairs and separated species are in equilibrium in THF.

7.4. Dinuclear Metal Carbonyls

$Mn_2(CO)_{10}$, $Re_2(CO)_{10}$, and Related Species

Data for $Mn_2(CO)_{10}$ and related species are given in Table 7.10.

The investigation of dimanganese decarbonyl has presented a number of problems, which are shared by the rhenium analogue. In both cases, the idealised site symmetry of the molecule in the crystal is D_{4d}, although the actual site symmetry is lower (C_2). There are two molecules in the primitive cell, and the isomorphous point group is C_2 [150]. There is every reason to suppose that the solution structures correspond to the idealised solid state structure. The situation is then as shown in Fig. 3.36, and the secular equations are given in Eqs (3.128, 3.131). Two different kinds of carbonyl group are present, and the energy factored force field contains nine parameters. There are seven carbonyl modes in $Mn_2(^{12}CO)_{10}$, of which three are infrared allowed and the remaining four are Raman allowed. In solution, none of the frequencies are allowed in both sets of spectra. Two of the solution Raman bands correspond to fully symmetric modes, and are therefore allowed to be polarised. In fact, only one band is found to be polarised, and there is no way in which the remaining bands can be assigned experimentally from the solution spectrum alone.

Four methods have been employed to solve the problems mentioned. These are, in historical order, the use of infrared combination spectra [96, 97], the use of data for regularly substituted species [143], the use of spectra for ^{13}CO-containing species [147] and the use of solid state Raman spectra [122]. Combination spectra were of considerable historical importance in establishing that interaction parameters linking CO groups on different metals are not neglible. These spectra, however, cannot be unambiguously assigned in the absence of other data, and, even if the correct assignment is chosen, lead to underestimates of the infrared-inactive frequencies. Data for PF_3-substituted species [134] were interpreted on the assumption that the energy factored force field is invariant on replacement of CO by PF_3, and that the interaction constant h of Fig. 3.36 is zero. The force field derived in this way implies values for the infrared-inactive frequencies that do not agree well with the observed Raman frequencies. These frequencies have been predicted more successfully [147] from the spectrum of ^{13}CO-containing species, using data superior to those available to earlier workers [134]. The calculation is based on the independent parameter method but the ^{13}CO data are used to define only one parameter in addition to the Raman frequencies, and the treatment appears to assume that the degree of mixing of axial and equatorial symmetry coordinates is the same in the A_1 and B_2 modes. This would be a reasonable assumption if the interaction parameter d that causes this mixing were far greater than the parameters linking CO groups on different metals; but this is apparently not so. Nonetheless, there is satisfactory agreement with the results of the single-crystal Raman study [122], which is totally unambiguous.†

The single crystal Raman study uses the "oriented gas" model of solids. Factor group analysis is not sufficient to distinguish between the E_2 and E_3 modes, both of which correlate with $a + b$ in the site group and hence with $A_g + A_u + B_g + B_u$ in the isomorphous point-group. However, the intensity distribution expected for the two modes is quite different. The E_2 modes transform as $xy, x^2 - y^2$; the former component belongs to B_g in C_{2h}, while the latter belongs to A_g. Similarly, the E_3 modes transform as xz (A_g), yz (B_g).‡ If light is initially polarised along the molecular x-axis, Raman scattering by the A_g mode correlating with E_2 will leave its direction of polarisation unchanged, while if the scattering is by the A_g component of E_3, the electric vector of the scattered light will correspond to the molecular z-axis. The usefulness of the oriented gas model is confirmed by the weakness or absence of Raman scattering from the infrared active modes of the isolated molecule, even though such scattering becomes formally allowed in the crystal, and by the consistency of the assignments derived.

The assignment of the infrared spectrum of $Mn_2(CO)_{10}$ has been confirmed by nematic solution studies [15]. In this case the assignment was not in

---

† Unfortunely, a fuller analysis [358] serves only to emphasise the difficulties of determining two parameters for one compound.

‡ The S_8 axis of D_{4d} is taken as z. These labels differ from those of [122], which are chosen in accord with the conventions for the isomorphous point group.

serious doubt, and the result therefore serves mainly to validate the nematic solution method. The intensity distribution in $Mn_2(CO)_{10}$ is most unusual, the higher frequency B_2 band being rather more intense than that at lower frequencies. ($Tc_2(CO)_{10}$ and $Re_2(CO)_{10}$ [148] do not show this abnormality.) This is unexpected, since the B_2 modes are closely related to the A_1 modes of species of the type $M(CO)_5L$ discussed in Chapter 3. The simplest explanation is that there is intense mixing of the axial and equatorial B_2 motions, with the equatorial motion being in this case the main contributor to the lower frequency vibration. This result can be reconciled with the usual assumptions ($k(a)$; c, t positive) if the parameter p (linking cisoid equatorial groups on different metals) is slightly larger, and j (linking the two axial groups) slightly smaller, than is found in the calculation of [147]. Whether the suggestion is correct or not, it is clear that the long-standing problem of the forms of the carbonyl modes in this molecule is still not fully resolved.

The infrared spectrum of $MnRe(CO)_{10}$ is very similar to those of $Mn_2(CO)_{10}$ and $Re_2(CO)_{10}$, except for the existence of a weak high frequency band [148]. This may with hindsight be equated with the highest A_1 band in the Raman spectra of $Mn_2(CO)_{10}$ and $Re_2(CO)_{10}$. In the unmixed carbonyls the local oscillating dipoles of the two halves of the molecule, moving in phase, are required by symmetry to cancel each other out, while in the mixed carbonyl the cancellation is not quite complete.

The anions $M_2(CO)_{10}{}^{2-}$ (M = Cr, Mo, W) are isoelectronic and isostructural [151] with $Mn_2(CO)_{10}$. The spectra are similar, but displaced to low frequency, and broadened. As far as can be seen, the lower B_2 band is less intense in the infrared spectrum than is the higher band. It seems likely that the explanation for this effect in $Mn_2(CO)_{10}$ holds good for these species as well.

Enneacarbonyldiiron, Octacarbonyldicobalt and Related Species

Frequencies for $Fe_2(CO)_9$, $Co_2(CO)_8$ and related species are presented in Table 7.11.

Data have been obtained for $Fe_2(CO)_9$ as a solid [152, 153] and in an argon matrix [78]. The data are consistent with the accepted [53] (D_{3h}) structure (Fig. 7.1), with moderate interaction between CO groups on different metals.

$Co_2(CO)_8$ has been shown [154, 155] to exist in solution as an equilibrium mixture of two isomers. One of these clearly corresponds to the solid, which is known to possess a bridged structure (Fig. 7.2). The other form contains no bridging CO groups, and would be expected, by analogy with species of the type $[LCo(CO)_3]_2$ [57], to possess a structure of D_{3d} symmetry. The observed infrared spectrum is consistent with such a structure, although at first

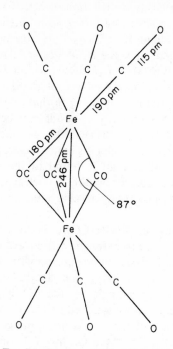

FIG. 7.1. Structure of Fe₂(CO)₉ [53].

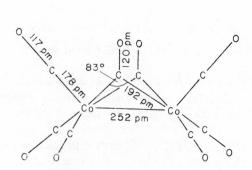

FIG. 7.2. Solid State Structure of Co₂(CO)₅ [62].

glance it may seem to preclude it. Three bands are predicted, and five are observed. The lowest frequency band can be dismissed as a ^{13}CO satellite. The highest band is very weak, and may be assigned to the highest A_{1g} mode, which is formally forbidden. The separation between the highest A_{1g} and A_{2u} modes would then be 37 cm^{-1}, which is reasonable and indicates a moderately strong interaction between CO groups on different metals. Such a weak high energy band is also observed in $Cd[Co(CO)_4]_2$ and $Hg[Co(CO)_4]_2$, for which some kind of D_3 structure is expected in solution, and its presence in these species confirms the assignment proposed for $Co_2(CO)_8$. The separations between the highest band and the highest formally allowed

TABLE 7.11. CO frequencies for $Fe_2(CO)_9$, $Co_2(CO)_8$ and related species.

Species	Solvent	Frequencies	Reference	Comments
$Fe_2(CO)_9$	Ar	2066, 2038 (A_2'', E' terminal); 1855, 1851 (E' bridging)	[78]	(a)
$Co_2(CO)_8$ (bridged)	H	2112w, 2071s, 2042s, 2001w (terminal); 1863w, 1853s (bridging).	[155]	
(non-bridged)	H	2106w(A_{1g}), 2069s (A_{2u}), 2031s (A_{2u}), 2022s(E_u), 1991m (^{13}CO).	[155]	
$Cd[Co(CO)_4]_2$	H	2081w(A_{1g}), 2071s (A_{2u}), 2016m (A_{2u}) 1995s (E_u) 1953w(^{13}CO)	[156]	(a)
$Hg[Co(CO)_4]_2$	H	2091vw(A_{1g}), 2072s (A_{2u}), 2021ms(A_{2u}), 2007vs(E_u), 1964w (^{13}CO)	[156]	(a)
	H	2094·6(A_{1g}), 2027·5 (A_{1g}), 2072·3 (A_{2u}), 2021·7(A_{2u}), 1996·0 (E_g), 2007·3 (E_u)	[157]	(b)

(a) Medium frequency data also given.
(b) Italicised frequencies inferred from data for ^{13}CO-containing species. The procedure may be criticised on the same grounds here as in its application to $Mn_2(CO)_{10}$ (see text), but the errors are less likely to be serious.

band in the infrared spectra of the Cd and Hg derivatives are $10\,cm^{-1}$ and $19\,cm^{-1}$ respectively, indicating an interaction between CO groups on different cobalt atoms smaller than that in $Co_2(CO)_8$, but still far from negligible. The size of the separation in $Hg[Co(CO)_4]_2$, compared with $Cd[Co(CO)_4]_2$, suggests an electronic coupling mechanism rather than an exciton mechanism, although a more trivial explanation in terms of differences in axial-equatorial couplings cannot be ruled out.

7.5. Substituted and Polynuclear Metal Carbonyls

Selected data for substituted carbonyls are given in Tables 7.12 to 7.16. These tables are arranged in order of increasing number of carbonyl groups. For a given number of carbonyl groups, the ordering is by increasing coordination number at the metal, with further ordering according to the periodic table; however, where convenient, arene and cyclopentadienyl complexes are grouped together. A selection of data for substituted dinuclear carbonyls and polynuclear carbonyls is given in Table 7.17. Frequencies outside the $v(CO)$ region are not quoted, for reasons of space, but are generally cited in footnotes. Assignments, where offered, are usually either straightforward or based on the views of the original authors, and attention is drawn to cases where this is not so.

Included in the tables are frequencies collected as part of detailed spectroscopic studies, together with data collected for routine characterisation; thus the limits of error vary from case to case. It also seems reasonable to suppose that the older work is on the whole less accurate. A wide range of solvents has been used, and attention is once again drawn to the danger of comparing frequencies obtained in different solvents. This is particularly true for solvents of different degrees of polarity, and for frequencies arising from CO groups *trans* to substituents.

The interpretation of the data is discussed in Chapter 6 above. On the whole, CO frequencies may be seen to correlate with electron availability at the metal, and to reflect the electronic influence of the other ligands, but there are cases where the ranking of substituent influences seems to vary from one series to another. Attention is also drawn to the effects of position in the Periodic Table, and to the undoubted reality of interaction constants involving two different metal sites, and of interactions involving bridging CO. For the rest, the data speak for themselves.

TABLE 7.12. CO Frequencies of some monocarbonyls.

Species	Solvent*	Frequency	Reference	Comments
NiF_2CO	Ar matrix	2200	[356]	
$Cu(en)CO^+Cl^-$	M	2080	[158]	
$[Cu(en)CO]_2Cl_2$	M	1905	[158]	CO-bridged dimer.
$HB(pyrazolyl)_3CuCO$	H	2083	[358]	
$(C_2H_5CCC_2H_5)_3WCO$	H	2034	[359]	
$Mn(NO)_3CO$	H	2088	[160]	(a)
$Fe(NO)_2PCy_3CO$	CCl_4	1997	[161]	(a, b)
$Fe(NO)_2P(C_6H_5)_3CO$	CCl_4	2010	[161]	(a)
$Fe(NO)_2As(C_6H_5)_3CO$	CCl_4	2011·5	[161]	(a)
$Fe(NO)_2Sb(C_6H_5)_3CO$	CCl_4	2011	[161]	(a)
$Fe(NO)_2P(OC_6H_5)_3CO$	CCl_4	2029·5	[161]	(a)
$CoNO[PCy_3]_2CO$	CCl_4	1927·5	[161]	(a, b)
$CoNO[P(C_6H_5)_3]_2CO$	CCl_4	1957	[161]	(a)
$CoNO[As(C_6H_5)_3]_2CO$	CCl_4	1957·5	[161]	(a)
$CoNO[P(OC_6H_5)_3]_2CO$	CCl_4	2004	[161]	(a)
$K_2[Co(NO)(CN)_2CO]$	(KBr)	1903	[162]	(a, c, d)
$Co(PF_3)_3CO^-$	acetone	1953	[163]	
$Ni(PF_3)_3CO$	H	2073	[164]	(c)
$Ni(PCl_3)_3CO$	H	2059	[165]	
$Ni(PCl_2Ph)_3CO$	H	2018·5	[165]	
$Ni(PClPh_2)_3CO$	H	1975·7	[165]	
$Ni(PCl_2OBu)_3CO$	H	2037	[165]	
$Ni[PCl(OBu)_2]_3CO$	H	2004	[165]	
$Ni[P(OBu)_3]_3CO$	H	1954	[165]	
$Ni(CNPh)_3CO$	$CHCl_3$	1971	[166]	(d)
$Ni(PMe_3)_3CO$	H	1923	[167]	(c)
$Ni(CNEt)_3CO$	H	1950	[168]	(d)

TABLE 7.12—contd.

Species	Solvent*	Frequency	Reference	Comments
Pd(PPh$_3$)$_3$CO	(solid)	1955	[169]	(e)
h^5–CpCuCO	H	2093	[170]	(f)
[Rh(CO)(PPh$_3$)Cl]$_2$	M	1982	[171]	(f)
[Rh(CO)(PPh$_3$)Br]$_2$	M	1977	[171]	(f)
[Rh(CO)(PPh$_3$)I]$_2$	M	1972	[171]	
[Rh(CO)ClPPh$_2$]$_2$	(KBr)	1999	[172]	Transoid, (g)
[Rh(CO)ClPPh$_2$]$_2$	(KBr)	2074, 1991	[172]	cisoid, (g)
[Ir(CO)ClPPh$_2$]$_2$	(KBr)	2042	[172]	(g)
[Ir(CO)ClAsPh$_2$]$_2$	(KBr)	2044	[172]	(g)
trans [RhA(CO)(PPh$_3$)$_2$]:				
A = CN	CHCl$_3$	2003	[173]	
NCSe	CHCl$_3$	2002	[173]	
NO$_2$	CHCl$_3$	1996	[173]	
N(CN)$_2$	CHCl$_3$	1995	[173]	
OClO$_3$	C$_6$H$_6$	1992	[173]	
NCS	CHCl$_3$	1990	[173]	
ONO$_2$	CHCl$_3$	1985	[173]	
SePh	CHCl$_3$	1982	[173]	
NCO	CHCl$_3$	1982	[173]	
I	CHCl$_3$	1981	[173]	
Br	CHCl$_3$	1980	[173]	
Cl	CHCl$_3$	1980	[173]	
N$_3$	CHCl$_3$	1980	[173]	
Sph	CHCl$_3$	1980	[173]	
O.CO.H	CHCl$_3$	1979	[173]	
O.CO.CH$_3$	CHCl$_3$	1972	[173]	
F	CHCl$_3$	1971	[173]	

Compound	Solvent	ν	Ref	Notes
OH	$CHCl_3$	1961	[173]	
$trans-[IrA(CO)(PPh_3)_2]$:				
A = CN	$CHCl_3$	1990	[173]	
NCSe	$CHCl_3$	1987	[173]	
NO_2	$CHCl_3$	1987	[173]	
$OClO_3$	C_6H_6	1982	[173]	
NCS	$CHCl_3$	1976	[173]	
NCO	$CHCl_3$	1968	[173]	
$C \equiv C - CH_3$	M	1968	[174]	
$C \equiv C - Bu^t$	M	1962	[175]	
I	$CHCl_3$	1967	[173]	
Br	$CHCl_3$	1966	[173]	
Cl	$CHCl_3$	1965	[173]	
F	$CHCl_3$	1957	[173]	
$O.SO.p-$tolyl	M	1955	[176]	
OH	$CHCl_3$	1949	[173]	
$dipyPtCl(CO)^+$		2145	[177]	various salts
$dipyPtBr(CO)^+$		2132	[177]	
$dipyPtI(CO)$		2120	[177]	
p-tolyl.$NH_2PtCl_2(CO)$		2136	[177]	
p-tolyl.$NH_2PtBr_2(CO)$		2129	[177]	
p-tolyl.$NH_2PtI_2(CO)$		2120	[177]	
$Ph_3P.PtCl_2(CO)$		2135	[177]	
$Ph_3P.PtBr_2(CO)$		2130	[177]	
$Ph_3P.PtI_2(CO)$		2103	[177]	
Cl_3PtCO^-		2106	[178]	
Br_3PtCO^-	M	2126	[177]	Halogens assigned as as mutually *trans*.
I_3PtCO^-	M	2096	[177]	
$[PtCl_2CO]_2$	M	2076	[177]	(c)
$[PtBr_2CO]_2$	M	2146	[177]	
	M	2135	[177]	

TABLE 7.12—contd.

Species	Solvent*	Frequency	Reference	Comments
[Pt₂CO]₂	M	2122	[177]	(f)
Fe(PF₃)₄CO— C₂ᵥ	liq.	2040	[359]	
Fe(PF₃)₄CO— C₃ᵥ	liq.	2011	[359]	
Fe(NO) C₃H₅)P(OMe)₃CO	toluene	1944	[179]	(a, h)
Fe(NO) (C₃H₅)PPh₃CO	toluene	1935	[179]	(a, h)
Fe(NO) (C₃H₅)AsPh₃CO	toluene	1929	[179]	(a, h)
Fe(C₄H₆)₂CO	cyclohexane	1984·5	[180]	(i)
RuH(OClO₃) (CNR)CO	M	1947	[181]	(j, k)
Ru(O₂) CNR) PPh₃)₂CO	M	1920	[181]	(j, k)
Ru(CNR) (PPh₃)₃CO	M	1901	[181]	(j, k)
RuCl(PPh₃)₂(CH₃CO)CO	M	1945	[182]	(l, m)
OsHCl(PCy₃)₂CO	(KBr)	1932, 1887	[183]	(b, n)
CoCl₂(PEt₃)₂CO	H	1977·1	[24]	(o)
Co(CN)₂(PPh₃)₂CO⁻		1918	[184]	
Co(CN)₂(PMe₂Ph)₂CO⁻		1875	[184]	
CoH(PF₃)₃CO	(vapour)	2076	[163]	
[RhI₄CO]⁻	(M; KBr)	2070	[185]	
Ir(SO₂.ptolyl) (O₂) (PPh₃)₂CO	M	2025	[176]	(p)
Ir(C ≡ CMe) (PPh₃)₂(Y) CO):				
Y = O₂	M	1998	[174]	
MeO₂C.C ≡ C.CO₂Me	M	2004	[174]	
SO₂	M	2011	[174]	
(CF₃)₂CO	M	2030	[174]	
C₄F₆	M	2041	[174]	
Ir(C₄H₆) (PMe₂Ph)₂ (CO)⁺	CHCl₃	2020	[186]	(q)
Ir(dppe)₂CO⁺Cl⁻	(KBr)	1929	[172]	(i)
Ni(fdma)I₂CO	C₆H₆	2055	[187]	(r), (s)

Compound	State	ν (cm⁻¹)	Ref.	Notes
Ni(h^5-Cp)(SiCl₃)CO	H	2062	[188]	*(a)*
π-CpRe(CH₃)(NO)(CO)	H	1972	[360]	
Mo(PF₃)₅CO	H	2029	[352]	
Mo(PMe₃)₅CO	H	1773	[354]	
Fe(CN)₅CO³⁻	Na salt	2040	[189]	
[Fe(CN)₅CO]₂Co₃	solid	1950	[189]	
RuHCl(CNR)(PPh₃)₂CO	M	1935, 1922	[181]	*(t)*
Ru(CO₃)(CNR)(PPh₃)₂CO	M	1950	[181]	*(j, k)*
[RuH(CNR)(PPh₃)₃CO]⁺	M	1990	[181]	*(j, k)*
OsHCl(PPh₃)₃CO	M	1912	[98]	*(k, u)*
OsD₂(PPh₃)₃CO	M	1918	[98]	*(k, v)*
Rh(PPh₃)₂Cl(C₂H₄)CO	CH₂Cl₂	2010	[172]	
IrHCl₂(PPh₃)₂CO	M	2027	[98]	*(k, u)*
IrD₂Cl(PPh₃)₂CO	M	2003	[98]	*(k, v)*
Ir(CS)PCy₃)₂CO	CHCl₃	2058	[190]	*(b, k, v)*
Ir(C ≡ CMe) (SO₄) (PPh₃)₂CO	M	2049	[174]	
Ir(C ≡ CBuᵗ) (CH₃)I(PPh₃)₂CO	M	2041	[175]	
Ir(Br)₂(CH₂Br) (PMe₂Ph)₂CO	CHCl₃	2037	[186]	
IrCl₃(PMe₂Ph)₂CO	CHCl₃	2073	[190]	CO *trans* to Cl
		2105	[190]	CO *trans* to phosphine
CpMn(PPh₃)₂CO	M	1836	[191]	*(e)*
CpMn(AsPh₃)₂CO	M	1836	[191]	*(e)*
CpMn(SbPh₃)₂CO	M	1825	[191]	*(e)*
CpFe(dmpe)CO⁺	CHCl₃	1960	[192]	*(w)*
(h^5Cp)Fe(PPh₃) (SnPh₃)CO	H	1923, 1911	[193]	*(e, x)*
CpFe(PPh₃) (SnMe₃)CO	H	1921, 1911	[193]	*(e, x)*
CpFe(PPh₃) (SnCl₃)CO	H	1969	[191]	
CpFePCy₃,COCMe(CO)⁺		1905	[194]	*(b, e)*
CpFePCy₃C(OMe)₂CO)⁺		1995	[194]	*(b, e)*
CpRuPCy₃[C(OMe) (OEt)]CO		1968	[194]	*(b, e)*
CpCo(SnMe₃)₂CO	H	1963	[195]	*(b, e)*
CpCo(CF₃)ICO	CS₂	2073	[196]	*(b, e)*

TABLE 7.12—contd.

Species	Solvent*	Frequency	Reference	Comments
CpCoH(SiCl$_3$)CO	H	2045	[197]	(b)
CpRh(X) (Y)CO:				
(X) (Y) = (H) (SiPh$_3$)	H	2024		
(SiCl$_3$)$_2$	H	2073		
(Cl) (GeCl$_3$)	CH$_2$Cl$_2$	2098	[198]	
(Br) (GeBr$_3$)	CH$_2$Cl$_2$	2092		
(I) (GeI$_3$)	CH$_2$Cl$_2$	2080		
(Cl) (SnCl$_3$)	CH$_2$Cl$_2$	2093		
C$_5$Me$_5$Ir(CO) (X) (Y):				
(X) (Y) = I$_2$	CH$_2$Cl$_2$	2044		
(CF$_3$) (I)	CH$_2$Cl$_2$	2050	[199]	
(CH$_3$) (I)	CH$_2$Cl$_2$	2002		
CpFeH(SiCl$_3$)$_2$CO	H	2025	[197]	(e)

(*) H = saturated hydrocarbon; M = mull.
(a) ν(NO) data also given.
(b) Cy = cyclohexyl.
(c) Medium frequency data also reported.
(d) ν(CN) data also given.
(e) Cp = C$_5$H$_5$.
(f) Halogen-bridged dimer.
(g) P–, As-bridged dimers. The separation of peaks claimed for cisoid Rh compound seems remarkably high, especially in the absence of a
     metal–metal bond.
(h) C$_3$H$_5$ = π-ally.
(i) C$_4$H$_6$ = h^4-butadiene.
(j) R not specified.

(k) stereochemistry discussed.
(l) frequency said not to vary significantly between nujol, KBr and benzene.
(m) ν(CO, acyl) = 1505 cm^{-1}. Compound described on this evidence as π–acyl.
(n) Different frequencies attributed to different isomers. Splitting also found in deuterated species, and in benzene solution.
(o) ν(^{13}CO) = 1932.6, ν(C^{18}O) = 1932.9. See Chapter 3.
(p) As [Ph$_4$As]$^+$ salt. [Bu$^n_4$N]$^+$ salts differ by up to 10 cm^{-1}.
(q) C$_4$F$_6$ = hexafluorobutadiene.
(r) dppe = Ph$_2$PC$_2$H$_4$PPh$_2$
(s) fdma = ferrocene-1, 1′-bisdimethylarsine.
(t) CO coordinated to cobalt via oxygen.
(u) CO *trans* to Cl. Frequency unaffected by deuteration.
(v) CO *trans* to H. Frequency shifted by deuteration. Quoted frequency is of deuterium derivative and is therefore presumably less perturbed by metal-hydrogen motions.
(w) dmpe = Me$_2$PCH$_2$CH$_2$PMe$_2$.
(x) frequencies corresponding to two rotamers.

TABLE 7.13. CO frequencies of some dicarbonyls

Species	Solvent (a)	Frequencies	(b)	Reference	Comments
$Mn(CO)_2NO$	Ar	1949	1967	[200]	(c, d)
$Fe(NO)_2(CO)_2$	H	2083·5	2035·1	[162]	(d, e)
	CCl₄	2084	2036·5	[161]	(d)
$Co(NO)(CO)_2L$:					
L = PCl₃	H	2073·1	2029·6	[201]	(d)
PPh₃	H	2035·0	1981·2	[201]	(f)
P(OPh)₃	CCl₄	2053·5	2001	[161]	(f, g)
AsPh₃	CCl₄	2038·5	1982·5	[161]	
PCy₃	CCl₄	2024	1966	[161]	
PEt₃	toluene	2032	1969	[202]	
Cl⁻	diglyme	2012	1938	[203]	
Br⁻	diglyme	2013	1940	[203]	
I⁻	diglyme	2011	1943	[203]	
CN⁻	diglyme	2019	1952	[203]	(h, i)
CN⁻	KBr	2030	1963	[162]	(h, i)
$Ni(CO)_2L_2$:					
L = PF₃	H	2094·0	2052·0	[164]	
PCl₃	H	2081·3	2044	[165]	
PCl₂(OBu)	H	2068·2	2023·5	[165]	
PCl(OBu)₂	H	2047·5	1995·5	[165]	
P(OBu)₃	H	2019	1963·5	[165]	
P(OPh)₃	H	2043	1996	[204]	
PCl₂Ph	H	2057·5	2017·5	[165]	
PClPh₂	H	2031·3	1979·5	[165]	
PPh₃	H	2005	1950	[165]	
P(OMe)₃	H	2027·2	1971·8	[167]	(e)
PMe₃	H	2001·5	1940·5	[167]	(e, j)

Compound	State/Solvent	$\nu(CO)$	Ref.	Notes
SbPh₃	H	2016, 1963	[205]	(h)
SbEt₃	H	2004, 1948·5	[205]	(k, l)
CNEt	H	2013, 1968	[168]	(k, l)
Rh(CO)₂Cl₂⁻	(KBr)	2060, 1975	[185]	(k, l)
Rh(CO)₂Br₂⁻	(KBr)	2055, 1975	[185]	(k, m)
Rh(CO)₂I₂⁻	(KBr)	2043, 1967	[185]	(k, m)
[Rh(CO)₂X]₂ : X≡Cl	CCl₄	2093s, 2038s, 1997m	[205]	(k, m)
	(KBr)	2087s, 2028s, 1863s,	[206]	
acetate	M	2083vs, 2067vs, 2027vs, 1980s	[207]	(k, m)
NCS	M	2083, 2033	[207]	
SO₄	M	2093vs, 2035s, 2014m	[207]	(k)
acacIr(CO)₂	H (solid)	2082, 2002	[208]	(d)
Cl₂Pt(CO)₂	x	2200, 2162	[177]	(d)
Mn(NO)(PPh₃)₂(CO)₂	CCl₄	1947·5, 1864	[161]	(d)
Mn(NO)(PCy₃)₂(CO)₂	CCl₄	1925, 1834	[161]	
Fe(NO)(C₃H₅)(CO)₂	Toluene	2034, 1975	[179]	
trans-Ru(PPh₃)₃(CO)₂	x	1905	[362]	
trans-Os(PPh₃)₃(CO)₂	x	1895	[362]	
CoCl(PPh₃)₂(CO)₂	CHCl₃	1984s, 1919vs	[209]	
CoBr(PPh₃)₂(CO)₂	CHCl₃	1988s, 1922vs	[209]	
CoI(PPh₃)₂(CO)₂	CHCl₃	1992s, 1928vs	[209]	
Co(CN)(PPh₃)₂(CO)₂	CHCl₃	1968s, 1932vs	[184]	
CoCp(CO)₂	(various)	2028, 1967	[210]	
Rh(C₆F₅)(PPh₃)₂(CO)₂	M	1978	[211]	(n)
Ir(SO₂.p-tolyl)(PPh₃)₂(CO)₂	M	2000, 1940	[176]	
Cr(CO)₂(phen)[P(OEt)₃]₂	CHCl₃	1827, 1743	[212]	
Mo(CO)₂(phen)[P(OEt)₃]₂	CHCl₃	1872, 1790	[212]	
W(CO)₂(phen)[P(OEt)₃]₂	CHCl₃	1825, 1740	[212]	
trans-Cr(CO)₂(dppe)₂	H	1840w, 1792s	[213]	
trans-Cr(CO)₂(dppe)₂⁺	M	1826–1840	[213]	
trans-Mo(PF₃)₄(CO)₂	H	1989	[134]	(o)

TABLE 7.13—contd.

Species	Solvent	Frequencies	(b)	Reference	Comments
cis-$Mo(PF_3)_4(CO)_2$	H	2048	2010	[134]	(h)
$K_4Cr(CN)_4(CO)_2$	(solid)	1738	1648	[214]	(h)
$K_3Mn(CN)_4(CO)_2$	(solid)	1917	1833	[214]	
cis-$Mo(CO)_2(diars)_2$	H	1887	1828	[215]	(p,q)
cis-$Mo(CO)_2(dppe)_2^+$	$CHCl_3$	1859	1785	[215]	
cis-$Mo(CO)_2(dppe)_2^{2+}$	CH_2Cl_2	1948	1892	[216]	
cis-$Mo(CO)_2[P(OMe)_3]_4$	CH_2Cl_2	1969	1907	[216]	
trans-$Mo(CO)_2(dppe)_2$	H	1909	1856	[217]	
	CH_2Cl_2		1878	[216]	
trans-$[Re(III)(CO)_2I_4]^-$	CH_2Cl_2	[2050 y]	1995	[363]	(r)
trans-$[Re(II)(CO)_2I_4]^{2-}$	CH_2Cl_2	[1915 y]	1875	[363]	(s)
$FeI(COCH_3)(CO)_2(PMe_3)_2$	H	2042vw	1961vs	[218]	(t)
$FeI(CH_3)(CO)_2(PMe_3)_2$	H	1998s	1940s	[218]	(t)
cis-$[RuH(CO)_2(CNR)(PPh_3)_2]^+$	M	2085s	2045s	[181]	
trans-$[RuH(CO)_2(CNR)(PPh_3)_2]^+$	M	2122w	2045s	[181]	
cis,cis,trans-$Ru(CO)_2I_2(PPh_3)_2$	C_6H_6	2055	1990	[364]	
all-trans-$Ru(CO)_2I_2(PPh_3)_2$	C_6H_6		2001	[364]	
$Os(CO)_2(PPh_3)_2Cl_2$	CCl_4	2064	1999	[215]	(q)
$Ru(CO)_2(PPh_3)_2(O.CO.H)_2$	CCl_4	2055	1994	[219]	
$Ir(CO)_2D_3PPh_3$	C_6H_6	2073	2055	[220]	(u)
$CpCr(CO)_2NO$	H	2025	1955	[221]	
$CpMo(CO)_2NO$	H	2020	1945	[221]	
$CpW(CO)_2NO$	H	2012	1933	[221]	
$CpMn(CO)_2L$:					
L = nothing	H	1955	1886	[77]	
AsF_3	H	2006	1953	[222]	
PF_3	H	1996	1938	[222]	(c)

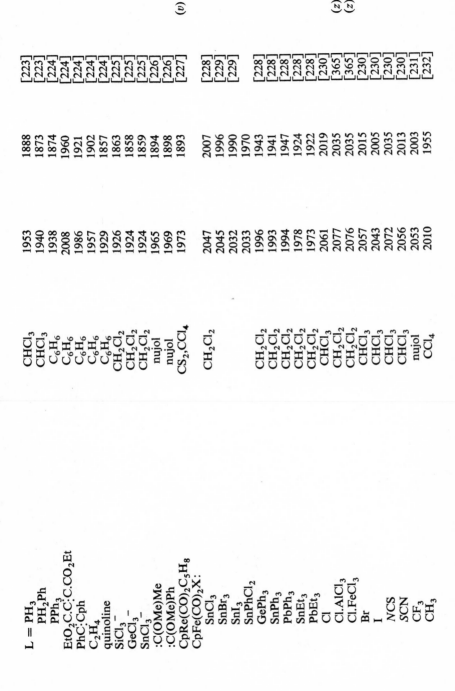

L =	solvent		(v)	(z)(z)
PH$_3$	CHCl$_3$	1953	1888	[223]
PH$_2$Ph	CHCl$_3$	1940	1873	[223]
PPh$_3$	C$_6$H$_6$	1938	1874	[224]
EtO$_2$C.C:C.CO$_2$Et	C$_6$H$_6$	2008	1960	[224]
PhC:Cph	C$_6$H$_6$	1986	1921	[224]
C$_2$H$_4$	C$_6$H$_6$	1957	1902	[224]
quinoline	C$_6$H$_6$	1929	1857	[224]
SiCl$_3^-$	CH$_2$Cl$_2$	1926	1863	[225]
GeCl$_3^-$	CH$_2$Cl$_2$	1924	1858	[225]
SnCl$_3^-$	CH$_2$Cl$_2$	1924	1859	[225]
:C(OMe)Me	nujol	1965	1894	[226]
:C(OMe)Ph	nujol	1969	1898	[226]
CpRe(CO)$_2$C$_5$H$_8$	CS$_2$,CCl$_4$	1973	1893	[227]
CpFe(CO)$_2$X:				
SnCl$_3$	CH$_2$Cl$_2$	2047	2007	[228]
SnBr$_3$		2045	1996	[229]
SnI$_3$		2032	1990	[229]
SnPhCl$_2$		2033	1970	
GePh$_3$	CH$_2$Cl$_2$	1996	1943	[228]
SnPh$_3$	CH$_2$Cl$_2$	1993	1941	[228]
PbPh$_3$	CH$_2$Cl$_2$	1994	1947	[228]
SnEt$_3$	CH$_2$Cl$_2$	1978	1924	[228]
PbEt$_3$	CH$_2$Cl$_2$	1973	1922	[228]
Cl	CHCl$_3$	2061	2019	[230]
Cl.AlCl$_3$	CH$_2$Cl$_2$	2077	2035	[365]
Cl.FeCl$_3$	CH$_2$Cl$_2$	2076	2035	[365]
Br	CHCl$_3$	2057	2015	[230]
I	CHCl$_3$	2043	2005	[230]
NCS	CHCl$_3$	2072	2035	[230]
SCN	CHCl$_3$	2056	2013	[230]
CF$_3$	nujol	2053	2003	[231]
CH$_3$	CCl$_4$	2010	1955	[232]

TABLE 7.13—contd.

Species	Solvent	Frequencies	(b)	Reference	Comments
n-C₃H₇	film	2013	1953	[233]	
C₆H₅	nujol	2021	1969	[231]	
COCH₃	C₆H₁₂	2035	1969	[234]	
COOCH₃	C₆H₁₂	2046	1995	[234]	
CH₂COCH₃	CS₂	2028	1976	[235]	(w)
SiCl₃⁺	C₆H₁₂	2039	1995	[188]	
PPh₃⁺	(nujol)	2070	2030	[236]	
AsPh₃⁺	(nujol)	2062	2017	[236]	
C₅Me₅Ir(CO)₂—σ-allyl⁺	CH₂Cl₂	2103	2063	[199]	
Cr(CO)₂(1, 3, 5-Me₃C₆H₃)L:					
L = quinnoline	C₆H₆	1873	1812	[233]	
PPh₃	C₆H₆	1886	1830	[233]	
C₂H₄	C₆H₆	1901	1852	[233]	
PhCCPh	C₆H₆	1923	1855	[233]	
CpV(CO)₂dppe	H	1855	1762	[237]	
CpV(CO)₂(PH₃)₂	H	2000	1964	[237]	
CpMn(H)(SiCl₃)(CO)₂	H	2028	1972	[377]	
Cp₂Ti(CO)₂⁺	C₆H₆,THF	1965	1885	[238]	
Cp₂V(CO)₂	M	2050	2010	[239]	

(a) Abbreviations of previous table used.
(b) s = strong, m = medium, w = weak, v = very.
(c) Photolytic fragment.
(d) NO frequencies also given.
(e) Medium frequency data also given.
(f) Data for species L = PCl₂Ph, PClPh₂ and for species Co(NO)(CO)L₂ also given.

(g) frequencies in CCl_4 or toluene similar [161, 202].

(h) CN frequencies also given.

(i) Note dramatic effect of phase.

(j) Typographical error in published paper corrected here (M. Bigorgne, private communication).

(k) CO groups mutually cis.

(l) As $[Ph_4As]^+$ salts. $[Bu_4N]^+$ salts differ by up to 10 cm^{-1}.

(m) Bridged dimers. Some interaction evident between $Rh(CO)_2$ units.

(n) Data for several related species also given.

(o) Frequency depends on nature of counterion.

(p) diars = o-phenylenebis (dimethylarsine).

(q) Frequencies in many other solvents compared and discussed.

(r) CO groups trans.

(s) CO groups cis.

(t) R not specified.

(u) fac arrangement of D atoms. Hydrogen analogue shows marked coupling of ν(Ir-H) and ν(CO).

(v) C_5H_8 = cyclopentene.

(w) Data for Me_6C_6 analogs also given.

(x) not stated.

(y) weak, in nujol, as $Bu^n_4N^+$ salt. Other frequencies in nujol are 1995[Ir(III)] and 1860[Ir(II)].

(z) coordination to metal trichloride is, no doubt, through a single chlorine bridge.

TABLE 7.14. CO frequencies of some tricarbonyls

Species	Solvent	$\nu(A_1)$	$\nu(E)$	Reference	Comments
Species M(CO)₃L					
Mn(CO)₃NO	Ar	2077, 1994, 1932		[200]	(a)
Co(CO)₃NO	(gas)	2108	2047	[162]	(b)
	H	2100	2033	[201]	(b)
	CCl₄	2099	2034	[161]	(b)
Ni(CO)₃L:					
L = PF₃	H	2110·5	2048·5	[164]	(c)
PCl₃	H	2103	2044	[165]	(d)
P(OMe)₃	H	2081·5	2014·5, 2005·5	[167]	
P(OMe)₃	H	2081·5	2014·5, 2005·5	[167]	
P(OBu)₃	H	2077	2008, 2000	[165]	
PMe₃	H	2068·5	1993·5	[167]	
SbEt₃	H	2067	1996	[205]	
SbPh₃	H	2074	2004·5	[205]	
BiEt₃	H	2072	1997·5	[205]	
CNEt	H	2071·5	2008	[168]	(d, f)
Species M(CO)₃L₂					
(i) Species of low symmetry					
Mn(NO) (CO)₃PPh₃	CCl₄	2034·5, 1971, 1926		[161]	(b)
Mn(NO) (CO)₃PCy₃	CCl₄	2029, 1959·5. 1910		[161]	(b)
C₄H₆Fe(CO)₃	H	2055·7, 1989·8, 1979·9		[240]	(g)
COTFe(CO)₃		2061, 1993, 1976		[241]	(h)
(π-C₃H₅) Fe(CO)₃	H	2046, 1968-60		[242]	(i)
Cl₃SnFe(CO)₃NO	H	2101, 2058, 2032		[243]	(b, d)

Ph$_3$GeFe(CO)$_3$NO	H	2073, 2020, 1990	[244]	(b)
Ph$_3$SnFe(CO)$_3$NO	H	2065, 2010, 1980	[243]	(b)
Ph$_3$PbFe(CO)$_3$NO	H	2060, 2006, 1978	[243]	(b)

(ii) Species of or approximating to 3-fold symmetry (k)

		A_1	E		
Fe(PMe$_3$)$_2$(CO)$_3$	C$_6$H$_{12}$	1954	1877	[139]	(l, m)
Fe(AsMe$_3$)$_2$(CO)$_3$	C$_6$H$_{12}$	1950	1875	[139]	(l, m)
Fe(SbMe$_3$)$_2$(CO)$_3$	C$_6$H$_{12}$		1880	[139]	(l, m)
Fe(PPh$_3$)$_2$(CO)$_3$	CHCl$_3$		1887	[245]	
Fe(CNMe)$_2$(CO)$_3$	CHCl$_3$	2009vw	1925	[245]	
Co(PPh$_3$)$_2$(CO)$_3^+$	CH$_2$Cl$_2$		2017	[209]	
	(KBr)	2077vw	2004	[209]	AlCl$_4^-$ salt
Co[P(OPh)$_3$]$_2$(CO)$_3^+$	CH$_2$Cl$_2$		2054	[209]	
Co(COC$_3$F$_7$) P(OPh)$_3$(CO)$_3$	CCl$_4$	2082w	2021, 2004	[246]	
CoC$_3$F$_7$P(OPh)$_3$(CO)$_3$	CCl$_4$	2086w	2014	[246]	
Co(SnBr$_3$)PPh$_3$(CO)$_3$	H	2056w	1999	[247]	
Co(H)PBu$_3$(CO)$_3$	H	2050w	1970 (1933w)	[248]	(n, o)
Ir(H)PBu$_3$(CO)$_3$	H	2039	1972 (1937w)	[248]	(o, p)

Species $M(CO)_3L_3$

(i) Species of low effective symmetry

trans-Mo(CO)$_3$[P(OEt)$_3$]$_3$	H	1988(A_1), 1890(A_1), 1883(B_1)	[249]	q
trans-Mo(CO)$_3$(PEt$_3$)$_3$	H	1954(A_1), 1841(A_1), 1846(B_1)	[249]	q
trans-Mo(CO)$_3$[P(OPh)$_3$]$_3$	H	2017(A_1), 1942(A_1), 1932(B_1)	[249]	q
trans-Cr[PMe(OMe)$_2$]$_3$(CO)$_3$	H	1972, 1887, 1871	[250]	
trans-W[PMe(OMe)$_2$]$_3$(CO)$_3$	H	1984, 1899, 1878	[250]	
trans-Mo(PF$_3$)$_3$(CO)$_3$	vapour	2072(A), 2022(A_1), 2003(B_1)	[251]	
trans-Mo(PF$_2$CF$_3$)$_3$(CO)$_3$	vapour	2081(A_1), 2036(A_1), 2015(B_1)	[251]	
Numerous mixed derivatives of Group VI metals			[212, 252, 253]	
Mn(PCy$_3$)$_2$(CO)$_3$CO$_2$Et	benzene	2033w, 1947vs, 1920vs	[254]	
Re(PCy$_3$)$_2$(CO)$_3$CO$_2$Me	benzene	2038w, 1943vs, 1920vs	[254]	
Fe(C$_3$H$_5$)(CO)$_3$Cl	H	2096s, 2051s, 2012s	[106]	(r)

TABLE 7.14—contd.

Species	Solvent	Frequencies		Reference	Comments
Fe(C₃H₅)(CO)₃Br	H	2089s, 2044s, 2011s		[106]	(r)
Ru(SiMe₃)(PPh₃)(CO)₃	H	2088w, 2022s, 1999m		[255]	(s)
Os(SiMe₃)(PPh₃)(CO)₃	H	2088w, 2014s, 1989m		[255]	
(ii) Species of or approximating to 3-fold symmetry (k)		A₁	E		
Mo(CO)₃L₃:					
L = PCl₃	CCl₄	2041	1989	[256]	
AsCl₃	H	2031	1992	[256]	
SbCl₃	H	2045	1991	[256]	
PPh₃	M	1949	1891	[256]	
AsPh₃	M	1957	1889	[256]	
AsEt₃	H	1938	1843	[257]	
SbPh₃	M	1972	1875	[256]	
CNPh	H	1953	1923-5	[166]	
PMe(OMe)₂	H	1970	1891	[250]	(t)
Mn(CO)₃(CN)₃²⁻	EtOH	2015	1937	[214]	
Mn(CO)₅dppeBr	CHCl₃	2022, 1959, 1918		[258]	
Mn(CO)₃dppeI	CHCl₃	2020, 1958, 1920		[258]	
Mn(CO)₃pteBr	CHCl₃	2037, 1966, 1931		[258]	
Mn(CO)₃pteI	CHCl₃	2032, 1963, 1932		[258]	(u)
Mn(CO)₃DiarsBr	CHCl₃	2022, 1956, 1913		[258]	
Mn(CO)₃DiarsI	CHCl₃	2018, 1953, 1915		[258]	
Mn(CO)₃(py)₂I	CHCl₃	2037, 1954, 1906		[259]	(v)
Mn(CO)₃dppeBr⁺	CH₂Cl₂	2104, 2069, 2016		[216]	
Re(CO)₃(py)₂I	CHCl₃	2041, 1934, 1891		[259]	
Re(CO)₃(PPh₃)₂I	CHCl₃	2049, 1996, 1904		[259]	
CpV(CO)₃²⁻	M	1748	1645	[260]	Cs salt

Compound	Solvent				Ref	Notes
$C_7H_7V(CO)_3^-$	CS_2	1980	1935sh, 1930		[261]	Na salt
$CpCr(CO)_3^-$	(KBr)	1876	1795		[260]	(w)
$C_6H_6Cr(CO)_3$	H	1987	1917		[262]	(w)
	C_6H_6	1975	1894		[262]	(w)
	C_6H_6	1946	1847		[262]	(w)
$C_5H_7NCr(CO)_3$	C_6H_6	1967	1887, 1873		[262]	
$C_4H_4SCr(CO)_3$	C_6H_6	1963	1888, 1869		[262]	
$C_4H_4SeCr(CO)_3$	H	1956-9	1883-7		[263]	
$C_6Me_6Cr(CO)_3$					[263]	
Numerous arene and triene complexes of $Cr(CO)_3$						
$CpMn(CO)_3$	C_6H_{12}	2028·1	1945·5		[263]	
$(C_6H_6CN)Mn(CO)_3$	H	2041	1969, 1963		[264]	CN exo
$CpRe(CO)_3$	CS_2, CCl_4	2041	1939		[226]	
$CpFe(CO)_3^+$	acetone	2121	2074		[265]	

Species $M(CO)_3L_4$ and $CpM(CO)_3L$

Compound	Solvent				Ref	Notes
$CpV(CO)_3L$:						
$L = PBu^n_3$	H	1951	1874	1854	[237]	
PH_3	H	1975	1901	1876	[237]	s
PPh_3	H	1962	1884	1865	[237]	
$CpMo(CO)_3X$:						
$X = BPh_2$	M	2058	1961	1901	[266]	
BCl_2	M	2049	1984	1949	[266]	
$SiCl_3$	H	2041	1976	1959	[188]	
$SnCl_3$	CH_2Cl_2	2049	1988	1964	[188]	
$Sn(CH_3)_2Cl$	CCl_4	2013	1947	1913	[267]	
$Sn(C_6H_5)_3$	CCl_4	2004	1934	1909	[267]	
$Sn(CH_3)_3$	CCl_4	1997	1922	1895	[267]	
Cl	$CHCl_3$	2063	1990br		[230]	(x, y)
Br	$CHCl_3$	2055	1983br		[230]	
I	$CHCl_3$	2040	1968 (1955sh)		[230]	
NCS	$CHCl_3$	2068	1997br		[230]	
SCN	$CHCl_3$	2056	1988, 1974		[230]	

TABLE 7.14—contd.

Species	Solvent	$\nu(A_1)$	$\nu(E)$	Reference	Comments
CF_3	M	2054	1976	[231]	
H	CS_2	2030	1949	[232]	
CH_3	CCl_4	2020	1937	[232]	
C_2H_5	CCl_4	2016	1932	[232]	
iso-C_3H_7	CCl_4	2010	1930	[232]	

(a) Photolytic fragment of low symmetry.
(b) NO stretching data also given.
(c) Medium frequency data also given.
(d) Data for related species also given.
(e) Note splitting of E bond, presumably by asymmetrical conformation of L.
(f) CN stretching frequency also given.
(g) ^{13}CO data, data for PF_3 derivatives, and energy factored force field analysis also given. $C_4H_6 =$ butadiene.
(h) COT = cyclooctatetraene. CS_2, CCl_4 and C_2Cl_4 used. Medium frequency region also discussed.
(i) Free radical, in equilibrium with dimer. Dimer absorbs at 2015, 1965 cm^{-1}.
(k) Labels given in C_3. Actual symmetries generally far higher.
(l) Medium and low frequency regions also analysed.
(m) Upper (A_1') frequency from Raman spectrum of solid.
(n) and references therein.
(o) Weak band at low frequency assigned (P.S.B.) as ^{13}CO satellite.
(p) Species unstable in absence of H_2, CO.
(q) Assignments revised and defended.
(r) Intensity data also given.
(s) Data given for numerous related species.
(t) Cr, W analogs at 1960, 1871 cm^{-1} and 1966, 1885 cm^{-1}.
(u) pte = 1, 2-bis(phenylthio)ethane, PhS.C_2H_4SPh.
(v) py = pyridine.
(w) C_5H_7N = N-methylpyrrole.
(x) br = broad
(y) Data for W analogs also given.

TABLE 7.15. CO frequencies of some tetracarbonyls

Species	Solvent	Frequencies				Reference	Comments
		A_1	A_1	A_1	E		
$CH_3COMn(CO)_4$	Ar	2087	2006		1957	[268]	(a)
$CH_3Mn(CO)_4$	(b)	2092	2002		1956	[268]	
$C_4H_6Fe(CO)_4$	C_2Cl_4	2084	2004		1981	[106]	(c, d)
$CH_3NCFe(CO)_4$	$CHCl_3$	2072	1996		1967	[269]	(e)
$Me_3PFe(CO)_4$	H	2051	1977		1936	[139]	(f)
$Me_3AsFe(CO)_4$	H	2049	1973		1938	[139]	(f)
$Me_3SbFe(CO)_4$	H	2046	1972		1938	[139]	(f)
$Ph_3PFe(CO)_4$	H	2052·2	1979·7		1946·6	[41]	(d)
$(OC)_4Fe-InBr_2{}^-$	CH_3CN		2012	1906		[270]	
$(OC)_4Fe-InBr_3{}^{2-}$	CH_3CN	1984	1890			[270]	
$HCo(CO)_4$	(gas)	2123	2062		2043	[271]	
$CH_3Co(CO)_4$	hexane	2116·1	2053·3		2029·8	[272]	(g)
	(gas)	2111·0	2045·9		2031·2	[272]	
	hexane	2104·6	2035·5		2018·5	[272]	(g)
$ClHgCo(CO)_4$	H	2099	2041		2018	[273]	
$BrHgCo(CO)_4$	H	2097	2040		2015	[273]	
$IHgCo(CO)_4$	H	2077·5	2015		1986·5	[273]	
$(OC)_4Co-CdBr_2 \cdot OEt_2{}^-$	CH_2Cl_2	2041	1996		1956	[270]	
$(OC)_4CoGaBr_3{}^-$	CH_2Cl_2	2084	2018		1991	[270]	
$(OC)_4CoInBr_3{}^-$	CH_2Cl_2	2080	2013		1989	[270]	
$(OC)_4CoSiCl_3$	H	2120·6	2064·8		2039·4	[41]	(d)
$(OC)_4CoSiPh_3$	H	2110·3	2033·5		2005·8	[41]	(d)
$(OC)_4CoGeBr_3$	H	2118	2066		2048	[274]	(h, i)
Numerous related $(OC)_4Co-M'$ species						[247, 275]	
$V(CO)_4(PPh_3)_2$	(KBr)		1850			[131]	E_u, trans
$V(CO)_4dppe^-$	THF	1903s, 1799vs, 1780s				[276]	

TABLE 7.15—*contd.*

Species	Solvent	Frequencies	Reference	Comments
Nb(CO)$_4$dppe$^-$	THF	1909s, 1803vs, 1780vs	[276]	
Ta(CO)$_4$dppe$^-$	THF	1908, 1800vs, 1779s	[276]	
cis (carbene)$_2$Cr(CO)$_4$	CHCl$_3$	1986, 1870, 1847, 1825sh	[277]	(*j*)
CF$_3$CO$_2$W(CO)$_4$PPh$_3$	CH$_2$Cl$_2$	2007, 1880sh, 1868s, 1812m	[278]	
W(CO)$_4$(DMF)$_2$	DMF	1997, 1858, 1832, 1791	[279]	(*k*)
Cr(CO)$_4$dipy	CHCl$_3$	2010m, 1908vs, 1888sh, 1833s	[280]	$A_1, B_1, A_1, B_2,$ (*l*)
Mo(CO)$_4$dipy	CHCl$_3$	2014m, 1908vs, 1892sh	[280]	
W(CO)$_4$dipy	CHCl$_3$	2008m, 1900vs, 1880sh	[280]	
Mo(CO)$_4$(CH$_3$)$_4$en	H	2014(A_1), 1888(A_1), 1881(B_1), 1856(B_2)	[249]	(*m, n*)
Mo(CO)$_4$en	CHCl$_3$	2015(A_1), 1864(A_1), 1890(B$_1$), 1818(B$_2$)	[249]	
cis-Cr[MeP(OMe)$_2$]$_2$(CO)$_4$	H	2022(A_1), 1934(A_1), 1911·5(B_1), 1909·5(B_2)	[250]	
cis-Mo[MeP(OMe$_2$]$_2$(CO)$_4$	H	2031·5(A_1), 1940·5(A_1), 1940·5?(B_1), 1921·0(B_2)	[250]	
cis-W[MeP(OMe)$_2$]$_2$(CO)$_4$	H	2028·5(A_1), 1943(A_1), 1934·5(B_1), 1912·5(B_2)	[250]	
trans-Cr[MeP(OMe)$_2$]$_2$(CO)$_4$	H	1883(E_u)	[250]	
trans-Mo[MeP(OMe)$_2$]$_2$(CO)$_4$	H	1919(?B_{2g}) 1894(E_u)	[250]	
trans-W[MeP(OMe)$_2$]$_2$(CO)$_4$	CHCl$_3$	1910(?B_{2g}) 1886(E_u)	[250]	
trans-Cr(PPh$_3$)$_2$(CO)$_4$	CHCl$_3$	2010vw(A_{1g}) 1945w(B_{2g}) 1889vs(E_u)	[34]	
trans-W(PPh$_3$)$_2$(CO)$_4$	CHCl$_3$	2005vw(A_{1g}) 1957w(B_{2g}) 1900vs(E_u)	[34]	
cis-Mo(CO)$_4$(AsR$_3$)$_2$; R = Et	H	2015, 1915, 1904, 1886	[257]	
Ph	H	2023, 1929, 1915, 1899	[257]	
OMe	H	2054, 1957, 1950, 1944	[257]	

Compound	Medium	Frequencies	Ref	Notes
trans-Mo(CO)₄(AsEt₃)₂	H	1888(E_u)	[257]	(o)
cis-Mo(CO)₄(SEt₂)₂	H	2021, 1909, 1904, 1874	[257]	
Cp₂Ti(SMe)₂Cr(CO)₄	CHCl₃	2000, 1908, 1892	[281]	(e)
SalenNiW(CO)₄	M	1899, 1790, 1667	[282]	(e)
dppePd(SMe)₂Cr(CO)₄	(nujol)	1978, 1856, 1849 1806	[283]	
cis-Mo(CO)₄(CNO)₂²⁻	THF	1994,1871, 1855sh,1806	[284]	
cis-Mo(CO)₄(CNEt)	H	2018(A₁), 1941·5(A₁), 1930(B₁), 1921·5(B₂)	[168]	(p)
[ClMn(CO)₄]₂	CCl₄	2104 2012 2045 1977	[285]	(q)
[BrMn(CO)₄]₂	CCl₄	2099 2011 2042 1975	[285]	(e)
[IMn(CO)₄]₂	CCl₄	2087 2009 2033 1976	[285]	(e)
[ClTc(CO)₄]₂	CCl₄	2119 2011 2048 1972	[285]	(e)
[ClRe(CO)₄]₂	CCl₄	2114 2000 2032 1959	[285]	(e)
[C₆F₅SMn(CO)₄]₂	H	2100 2020 2045 1995	[286]	(e)
[C₆F₅SRe(CO)₄]₂	H	2105 2000 2030 1964	[286]	(e)
[Ph₂PMn(CO)₄]₂	C₂Cl₄	2053 1992 1992 1957	[287]	
Re(CO)₄phen⁺	acetone	2123(A₁), 2032(A₁), 2010(B₁), 1967(B₂)	[254]	
Mn(CO)₄Cl₂⁻	C₂H₄Cl₂	2098, 1986 ,2026, 1936	[288]	
Mn(CO)₄Br₂⁻	C₂H₄Cl₂	2098, 1990, 2021, 1939	[288]	
Mn(CO)₄I₂⁻	C₂H₄Cl₂	2082, 1986, 2006, 1939	[288]	
Mn(CO)₄(CN)₂⁻	C₂H₄Cl₂	2103, 2005, 2029, 1936	[288]	
trans-Mn(CO)₄(PPh₃)₂⁺	THF	2001	[254]	
C₂F₅Mn(CO)₄P(OPh)₃	CCl₄	2103, 2029, 2016, 1988	[246]	
cis-Fe(CO)₄Cl₂	C₂Cl₄	2164(A₁), 2124(B₁), 2108(A₁), 2084(B₂)	[106]	(e)
cis-Fe(CO)₄Br₂	H	2148·5(A₁), 2106·7(B₁), 2097·2(A₁), 2073·8(B₂)	[30]	
cis-Fe(CO)₄I₂	H	2128·8(A₁), 2083·5(A₁, B₁), 2060·3(B₂)	[30]	
cis-Fe(CO)₄(SiCl₃)₂	H	2125, 2078, 2071, 2061	[188]	
Ru(CO)₄I₂	H	2161, 2106, 2079, 2068	[215]	

TABLE 7.15—contd.

Species	Solvent	Frequencies	Reference	Comments
Os(CO)$_4$Br$_2$	H	2177·8, 2089·8, 2112·8, 2051·0	[215]	
Os(CO)$_4$I$_2$	H	2169·8, 2085·8, 2100·8, 2050·0	[215]	
Os(CO)$_4$D$_2$	H	2140(A_1), 2057(A_1), 2049(B_1), 2031(B_2)	[289]	(r)
trans—Me$_3$SiRu(CO)$_4$Br	H	2058 (2027, ^{13}CO)	[235]	(e)
CpV(CO)$_4$	H	2031(A_1), 1931(E)	[237]	
C$_6$H$_6$V(CO)$_4^+$	H	2068(A_1), 2018(B_2), 1986(E)	[290]	
C$_6$Me$_6$V(CO)$_4^+$	H	2054(A_1), 1968(E)	[290]	
C$_6$H$_7$V(CO)$_4$	H	2021, 1963(mw), 1935(m), 1920(s)	[291]	

(a) Photolytic fragment.
(b) 'matrix mixture'.
(c) C$_4$H$_6$ = h^2-butadiene.
(d) Intensity data also given.
(e) Data for related species also given.
(f) Medium and low frequency regions fully analysed.
(g) ^{13}CO data analysed and energy factored force field determined.
(h) Full analysis of spectrum offered, for all frequencies.
(i) Laser Raman studies confirm assignments.

(j) carbene = 1,3-dimethyl-4-imidazolin-2-ylidene, CH$_3$N—CH$_2$ = CH$_2$—N(CH$_3$)—C: .
(k) DMF = dimethylformamide.
(l) Assignment (P.S.B.) from given intensities.
(m)CH$_4$ en = (N,N,N′,N′, tetramethy) ethylenediamine.
(n) Frequency order in CHCl$_3$ resembles following entry. Note dramatic effect of polar solvent on CO group trans to N only.
(o) SPh, and Cr, W, derivatives also described.
(p) CN stretching data also given.
(q) Interaction between CO group on different metals ignored. This is not strictly justified (see [Mn(CO)$_4$PMe$_2$]$_2$, Table 7.17).
(r) Frequencies for H derivative perturbed by coupling.

TABLE 7.16. CO frequencies of some pentacarbonyls

Species	Solvent	Frequencies				Reference	Comments
		A_1	B_2	E	A_1		
$[Cr(CO)_5]^-$	Ar matrix			1855	1838	[347]	(a)
$V(CO)_5NO$	H	2108		1992	2064	[131]	(b, c)
$V(CO)_5PPh_3^-$	THF	1965	1858	1823		[276]	(b)
$Nb(CO)_5PPh_3^-$	THF	1971	1863	1830		[276]	(b)
$Ta(CO)_5PPh_3^-$	THF	1973	1863	1828		[276]	(b)
$V(CO)_5AsPh_3^-$	THF	1969	1859	1826		[276]	(b)
$V(CO)_5PBu^n_3^-$	THF	1963	1856	1823		[276]	(b)
$V(CO)_5MeTHF^-$	MeTHF	1965	1820	1795	1758sh	[80]	(d)
$[V(CO)_5CN]^{2-}$	CH_3CN	1852		1793	1744	[356]	
$[V(CO)_5NH_3]^-$	M	1979		1787	1750	[366]	
$M(CO)_5X$ (M = Cr^{-1}, Mo^{-1}, W^{-1}, Mn, Re; X = Cl, Br, I)						[292]	(e)
$Cr(CO)_5I$	CH_2Cl_2	2100		2024	1988	[293]	
$Cr(CO)_5I^-$	CH_2Cl_2	2054		1920	1867	[293]	
$Cr(CO)_5CN$	$CHCl_3$	2049·5		1948·5	1983·5	[293]	(f)
$Cr(CO)_5CN^-$	$CHCl_3$	2052·5		1890	1930	[293]	
$Cr(CO)_5NCS^-$	$CHCl_3$	2115	2066	1934	1882	[294]	
$Mo(CO)_5NCS^-$	CH_2Cl_2	2105	2073	1941	1876	[294]	
$W(CO)_5NCS^-$	CH_2Cl_2	2105	2067	1920	1873	[294]	
$C_2F_5CO_2Cr(CO)_5^-$	CH_2Cl_2	2061		1918	1854	[278]	(g)
$C_2F_5CO_2W(CO)_5^-$	CH_2Cl_2	2068		1909	1835	[278]	(g)
$Mo(CO)_5(C_2H_4)$	H	2087		1960	1976	[295]	
$Mo(CO)_5(C_2H_2)$	H	2092		1974	1950	[295]	
$W(CO)_5(C_2H_4)$	H	2088		1953	1973	[295]	
$W(CO)_5(C_2H_2)$	H	2095		1967	1952	[295]	

SELECTED VIBRATIONAL DATA FOR CARBONYLS

TABLE 7.16—contd.

Species	Solvent	Frequencies A₁	B₂	E	A₁	Reference	Comments
Cr(CO)₅.COCH₃⁻	M	2041		1883		[226]	(h)
Mo(CO)₅.COCH₃⁻	M	2049		1883		[226]	(h)
W(CO)₅.COCH₃⁻	M	2058		1887		[226]	(h)
Cr(CO)₅C(CH₃)OEt	H	2064·1	1982·8	1946·9	1961·4	[296]	(i)
W(CO)₅C(CH₃)OEt	H	2066·5	1980·0	1944·8	1957·8	[296]	(d)
Mo(CO)₅CNEt	H	2069		1958·5	1958·5	[168]	
Cr(CO)₅MeTHF	MeTHF	2075		1935	1884	[45]	
Mo(CO)₅O(Pr)₂	OPr₂	2079		1940	1893	[279]	
W(CO)₅OEt₂	H	2074		1931	1908	[279]	
W(CO)₅CH₃CN	H	2083		1948	1931	[279]	
W(CO)₅DMF	DMF	2067		1917	1847	[279]	
W(CO)₅(Me₂CO)	Me₂CO	2067		1920	1847	[279]	
Cr(CO)₅NH₃	THF	2061w	1973w	1921vs	1879s	[297]	
Cr(CO)₅PH₃	H	2075m	1982sh	1953vs	1953vs	[297]	
Cr(CO)₅AsH₃	H	2072s	1982vw	1951vs	1951vs	[297]	
Cr(CO)₅SbH₃	H	2072s	1986w	1956vs	1956vs	[297]	
Mo(CO)₅SbH₃	H	2080s	1988w	1943vs	1943vs	[297]	
W(CO)₅SbH₃	H	2079s	1982w	1953vs	1953vs	[297]	
W(CO)₅Sb(C₅H₅)₃	H	2070m		1943vs	1943vs	[297]	
Cr(CO)₅MeP(OMe)₂	H	2071	1987	1947	1963·5	[250]	
Mo(CO)₅MeP(OMe)₂	H	2078·5	1989	1953	1967	[250]	
W(CO)₅MeP(OMe)₂	H	2077	1983	1947·5	1961	[250]	
Mo(CO)₅PCl₃	H	2094·5		1985	1999	[217]	
Mo(CO)₅P(OPh)₃	H	2083		1963	1975	[217]	

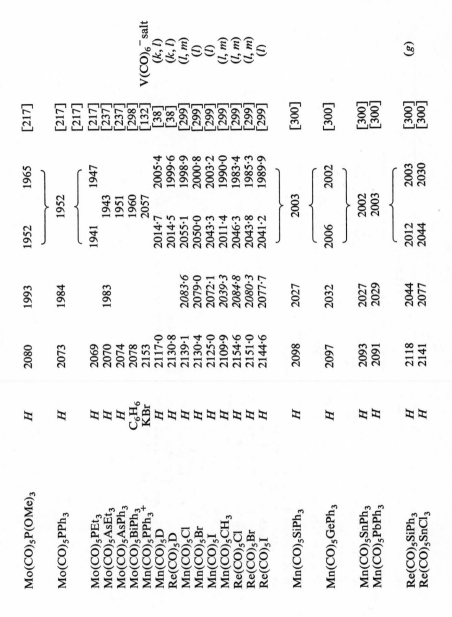

Compound	Solvent						Ref.	Notes
Mo(CO)₅P(OMe)₃	H	2080	1993	1952	1952	1965	[217]	
Mo(CO)₅PPh₃	H	2073	1984	1952	1952		[217]	
Mo(CO)₅PEt₃	H	2069	1983	1941	1952	1947	[217]	
Mo(CO)₅AsEt₃	H	2070		1943			[237]	
Mo(CO)₅AsPh₃	H	2074		1951			[237]	
Mo(CO)₅BiPh₃	C₆H₆	2078		1960			[298]	
Mn(CO)₅PPh₃⁺	KBr	2153		2057			[132]	V(CO)₆⁻ salt
Mn(CO)₅D	H	2117·0		2014·7	2005·4		[38]	(k, l)
Re(CO)₅D	H	2130·8		2014·5	1999·6		[38]	(k, l)
Mn(CO)₅Cl	H	2139·1	2083·6	2055·1	1998·9		[299]	(l, m)
Mn(CO)₅Br	H	2130·4	2079·0	2050·0	2000·8		[299]	(l)
Mn(CO)₅I	H	2125·0	2072·1	2043·3	2003·2		[299]	(l)
Mn(CO)₅CH₃	H	2109·9	2039·3	2011·4	1990·0		[299]	(l, m)
Re(CO)₅Cl	H	2154·6	2084·8	2046·3	1983·4		[299]	(l, m)
Re(CO)₅Br	H	2151·0	2080·3	2043·8	1985·3		[299]	(l, m)
Re(CO)₅I	H	2144·6	2077·7	2041·2	1989·9		[299]	(l)
Mn(CO)₅SiPh₃	H	2098	2027	2003	2003		[300]	
Mn(CO)₅GePh₃	H	2097	2032	2006	2002		[300]	
Mn(CO)₅SnPh₃	H	2093	2027	2002	2002		[300]	
Mn(CO)₅PbPh₃	H	2091	2029	2003	2003		[300]	
Re(CO)₅SiPh₃	H	2118	2044	2012	2003		[300]	
Re(CO)₅SnCl₃	H	2141	2077	2044	2030		[300]	(g)

TABLE 7.16—contd.

Species	Solvent	Frequencies				Reference	Comments
		A_1	B_2	E	A_1		
$Mn(CO)_5C_5F_4N$	H	2132	2068	2042	2015	[301]	(n)
$Mn(CO)_5SCN$	M	2138	2084	2043	1958	[302]	(i)
$Mn(CO)_5NCS$	MeCN	2141	2113	2053	1958	[302]	
$Mn(CO)_5SO_2Me$	CCl_4	2139	2090	2059, 2044	2027	[378]	

(a) Assignment based on idealised C_{4v} symmetry.
(b) Tentative assignment (P.S.B.).
(c) NO stretching data also given.
(d) MeTHF = 2-methyltetrahydrofuran.
(e) Medium frequency range only.
(f) Frequencies in CH_2Cl_2 very similar.
(g) Data for related species also given.
(h) As Me_4N^+ salt.
(i) CN stretching data also given.
(k) Frequencies in hydride perturbed by coupling.
(l) ^{13}CO data and energy factored force field also given.
(m) Italicised frequencies calculated from force field determined using ^{13}CO data.
(n) C_5F_4N = perfluoropyridyl.

TABLE 7.17. CO frequencies for some di- and polynuclear species

Species	Solvent	Frequencies	Reference	Comments
[Ph₃PMn(CO)₄]₂	CS_2	2055(A_1), 1983(B_2), 1972(E_2), 1954(E_1), 1913(E_3)	[67]	(a, b)
[Ph₃PRe(CO)₄]₂	CS_2	2066(A_1), 1991(B_2), 1979(E_2), 1952(E_1), 1915(E_3)	[303]	(a)
Hg[Co(CO)₄]₂	H	2094·6(A_{1g}), 2027·5(A_{1g}), 1966·0(E_g), 2072·3(A_{2u}), 2021·7(A_{2u}), 2007·3(E_u)	[157]	(c)
[Et₃PCo(CO)₄]₂	CS_2	2031·1vw(A_{1g}), 1969·9(A_{2u}), 1951·1(E_u), 1926·7(E_g)	[304]	
[Et₃PCo(CO)₄]₂Hg	CS_2	2013·5(A_{1g}), 1985·6(A_{2u}), 1943·1(E_u), 1932·3(E_g)	[304]	
[Fe(CO)₃PPh₂]₂	CCl_4	2049, 2010, 1979, 1963, 1953sh	[59]	
[Fe(CO)₃SEt]₂	H	2072, 2036, 2001, 1991; 2069, 2036, 1996, 1989	[305]	(d)
Co₂(CO)₆(C₆H₂)	H	2098·6mw, 2058·8s, 2034·0s, 2028·2ms, 2016·7w	[306]	(e, f)
(OC)₄Cr(PMe₂)₂Cr(CO)₄	CCl_4	2044, 2011, 1951, 1943sh	[59]	(e)
(OC)₄Mn(PMe₂)₂Mn(CO)₄	CCl_4	2065, 2042, 1979, 1957	[59]	
Cp₂Fe₂(CO)₄, transoid	H	1960; 1796	[307]	(g)
Cp₂Fe₂(CO)₄, cisoid	H	2004, 1960; 1812w, 1796	[307]	(g)
Cp₂Fe₂(CO)₄·AlBuⁱ₃	H	2065; 1985; 1828, 1682	[308]	(g, h)
Cp₂Fe₂(CO)₄·(AlBuⁱ₃)₂	H	2042; 2004; 1682	[308]	(h)
[CH₂: C(Me), C(Me): CH₂Co(CO)₂]₂	H	1977; 1811	[309]	(g, i)
[NBDCo(CO)₂]₂	H	2019, 1995; 1816sh, 1804	[309]	(g, j)
[CpMo(CO)₃]₂	CCl_4	2000, 1961, 1917, 1913(?), 1909	[59]	(e, k)
[CpNi(CO)]₂	H	1896, (1884), 1854, (1825)	[310]	(l)
CH₃CCo₃(CO)₉	H	2102, 2052, 2038, 2018	[311]	(a)

TABLE 7.17—contd.

Species	Solvent	Frequencies	Reference	Comments
SCo$_3$(CO)$_9$	H	2103·4, 2049·5, 2037·5, 2023·5	[312]	
Os$_3$(CO)$_{12}$	(solid)	2130(A_1', a_1), 2070(A_2'', b_1), 2028(E'', b_1), 2019(E',a_1); 2006(A_1', a_1), 2000(E', a_1), 1989(E',b_2), 1979(A_2', b_2)	[313]	(m, n)
RuFe$_2$(CO)$_{12}$	H	2115w, 2052s, 2040s, 2021sh, 2000m, 1989sh, 1860vw, 1828w	[313]	
Fe$_3$(CO)$_{12}$	H	2046s, 2023m, 2013sh, 1867vw, 1835w	[313]	
	Ar	2110vw, 2056s, 2051s, 2036s, 2032m, 2021w, 2013m, 2003vw, 1871sh, 1867w, 1833ms		
Co$_4$(CO)$_{12}$	H	2103w, 2063vs, 2055vs, 2037·5m, 2027m, 2018wsh, 1990vw; 1898w, 1866·5m, 1831w	[314]	(o)
[Mn(CO)$_3$]$_4$(SEt)$_4$	CCl$_4$	2044(A_1), 2016(T_2), 1954(E), 1942(T_2), 1923(T_1)	[106]	(p)
Os$_4$O$_4$(CO)$_{12}$	CHCl$_3$	2123(A_1), 2097(T_2), 2026(E), 2015(T_2)	[40]	(e)
[RU(CO)(π–C$_5$H$_5$)]$_4$	CHCl$_3$	1616	[367] [357]	(q)

(a) Data for related species also given.
(b) Data given in other solvents as well.
(c) Italicised frequencies inferred using ^{13}CO data.
(d) Two isomers.
(e) Italicised frequency estimated from binary spectrum; true value probably a few cm^{-1} higher in some cases.
(f) In cyclohexane, 2013, 1960sh, 1956, 1949.
(g) Lower frequency band(s) due to bridging CO.

(h) Band at 1682 due to bridging CO also coordinated through oxygen to Al.

(i) Bridged, assigned as pure transoid.

(j) Bridged, assigned as pure cisoid.

(k) See also Chapter 5.

(l) Assigned structure with puckered $Ni_2[CO]_2$ ring. Frequencies in brackets assigned as ^{13}CO satellites.

(m) Assignment based on Raman-infrared comparisons. Frequencies from 2006 cm^{-1} down are due mainly to the radial CO groups. Note that interaction between CO groups on different metals is strong enough to place higher E' (a_1) below A_2''.

(n) Data for Ru species also given.

(o) Structure in Ar matrix probably resembles that in solid, while solution structure does not.

(p) Bands below 1900 due to bridging CO, assignable as 1898(A_1), 1866-5(E), 1831(^{13}CO); 1990 cm^{-1} band could be ^{13}CO satellite of 2027 band.

(q) Assignment in T_d, Italicised frequencies from Raman spectrum.

Chapter 8

Non-Vibrational Spectroscopy of Carbonyls

This chapter is restricted to the photoelectron, electronic, and carbonyl ^{13}C nuclear magnetic resonance spectra of carbonyl complexes. Other spectroscopic methods, such as ^{1}H and ^{31}P n.m.r. spectroscopy, which give valuable complementary information about the non-carbonyl ligands and their interaction with the metal, lie outside the range of this book.

At the time of writing it appears that the apparatus required for photoelectron or for ^{13}C n.m.r. spectroscopy will never become as cheaply and readily available as are infrared instruments. There is, however, no underlying difficulty in giving at least a superficial interpretation of whatever information exists. The problem with electronic (visible – u.v.) spectroscopy is rather different. Apparatus is readily available for the range of greatest interest, but interpretation is controversial and difficult. For this reason, many chemists feel no need to publish electronic spectra; this is unfortunate, since the accumulation of data is a necessary step towards understanding.

8.1. Photoelectron Spectroscopy

Photoelectron spectroscopy is a technique in which a sample is bombarded with photons of known energy. This energy is chosen to be above the ionisation threshold of the sample, and the surplus energy is carried away by the ejected electrons. Corrections must of course be made for vibrational and rotational excitations of the sample. Electronic excitations of the sample can usually be described by specifying the nature of the electron removed. (Possible exceptions in the low bonding energy spectrum of $V(CO)_6$, and in the high bonding energy spectrum of chromium carbonyl complexes, are discussed below.) If E is the kinetic energy of the ejected electron, and ΔE(rot) and $\Delta(E$(vib)) are the changes in the rotational and vibrational energies of the sample, then energy conservation requires that

$$E = h\nu - \Delta E(\text{rot}) - \Delta E(\text{vib}) - I \qquad (8.1)$$

where I is the relevant ionisation potential. In any photoelectron experiment, the frequency of the exciting light is fixed, so that $h\nu$ is constant. The electrons ejected are passed through an energy analyser to a detector, and the spectrum is a plot of detector current against the energy collected. This is equivalent to a plot of the energies of the various ionisation processes possible, up to a limit imposed by the energy of the source radiation.

Two types of source are in use, producing "low" and "high" energies. The "low energy" source consists of a discharge through a noble gas. The most commonly used of these is helium, which generates lines at 584 nm (21·22eV) by the process He(Is2p) → He(Is^2), and at 303 nm (40·8eV) due to the process He$^+$(2p) → He$^+$(Is); the latter line can be suppressed or intensified by varying the operating conditions of the discharge. Low energy sources are preferred for the examination of valence electrons. The bandwidth of the exciting light is small, and since the kinetic energy of the ejected electrons is low, the resolution of the analyser is used to best advantage.† The "high energy" source is an x-ray tube, typically emitting Al Kα radiation (1487 eV). Radiation of this high energy is capable of removing "core" electrons, such as carbon and oxygen Is (binding energies around 289 and 536 eV) and the 3s, 3p electrons of complexes of the first row transition elements (with binding energies in the range 35–80e V). On the simplest view, the binding energies of core electrons reflect the net charge on the atom; the more positive this is, the more tightly the electrons are held. The usefulness of high-energy photoelectron spectroscopy (sometimes known as Electron Spectroscopy for Chemical Analysis, ESCA) is limited by various problems of resolution. Uncertainty broadening occurs in the source, so that the energy of the exciting light has a considerable spread, although it seems likely that this particular limitation will be overcome. The kinetic energies to be analysed are frequently large, so that the limiting resolution of the analyser corresponds to a wide range of energies. The last, and most intractable, problem is the uncertainty broadening of the very short-lived excited states generated by removal of a core electron. This imposes an ultimate limitation on the usefulness of the technique.

Low energy photoelectron spectra of molecules are generally collected in the vapour phase. There are both experimental and theoretical reasons for

† The uncertainty due to an analyser of resolution 1,000 is 5 meV for 5 eV electrons, but 1·4 eV for 1400 eV electrons.

this. In a molecular crystal, the valence electrons of the individual molecules interact to form broadened bands, and the energy of the electrons emitted from such a crystal is affected by the buildup of static charges. High energy photoelectron spectra, on the other hand, may be obtained for solids as well as for vapours; in fact, solids are generally preferred for reasons of intensity. The band structure broadening of core electrons is negligible. Errors due to sample potential can be minimised by calibration with a standard substance applied to the surface of the sample, and the poorness of the resolution obtainable reduces the importance of any errors that remain. Both low [9] and high [8] energy photoelectron spectroscopy have been authoritatively reviewed.

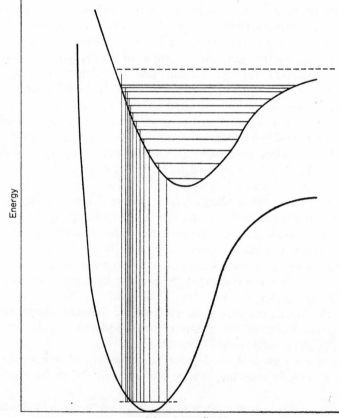

Internal coordinate

FIG. 8.1. Vibrational excitations broadening on electronic band where ground and excited state dimensions differ.

Low Energy Photoelectron Spectra of Carbonyls

The first ionisation potentials of metal carbonyls correspond in nearly all cases studied to the removal of an outermost d-electron; the energy required for this process is in the range 8–10 eV. Where more than one d-electron energy level is populated, ionisations from each level are observed, with integrated intensities proportional to their populations. In the vast majority of cases, the ionisation corresponds to removal of an electron from a closed shell. This process gives rise to one peak only for each level. A notable exception is vanadium hexacarbonyl. This has the open shell configuration $(t_{2g})^5$, and and ionisation can generate the configuration $(t_{2g})^4$ in any of its different spectroscopic terms. As a result, the first photoelectron band is split into two components at least one of which is believed to possess unresolved complexities [315].

The energy required to remove an electron is commonly taken as a measure of the bonding energy of that electron. Strictly speaking, this is an approximation. The removal of one electron is accompanied by a certain degree of redistribution of all the others so as to minimise the energy of the ion formed. It follows, after correcting for rotational and vibrational changes, that the observed ionisation potential is less than that predicted by a strict one-electron model of the ionisation process.

Although the spectra of carbonyl complexes have generally been collected in the vapour phase, the vibrational and rotational structure has not proved resolvable. However, the bands due to the loss of a d-electron are all moderately broad (halfwidth around 0·2 eV), as required by the Franck–Condon principle [316] if the electron lost has some considerable bonding (or antibonding) character (Fig. 8.1). The binding energies quoted here are for the "vertical" ionisation, in which the internuclear distances are unchanged. The bands due to removal of an electron from coordinated CO occur over a broad range of binding energies, from about 13 eV upwards, and are incompletely resolved.

The spectrum of vanadium hexacarbonyl has already been discussed; the observed ionisation energies are 7·52 eV (assigned as $^2T_{2g} \rightarrow {}^3T_{1g}$) and 7·88 eV (assigned as $^2T_{2g} \rightarrow {}^1T_{2g}$, $^2T_{2g} \rightarrow {}^1E_g$: the $^2T_{2g} \rightarrow {}^1A_{1g}$ is probably concealed beneath this second band). First order calculations show that the small separation of the observed bands corresponds to an extremely high nephelauxetic effect (see Section 8.2 below) for CO, but this result is thought to be an artefact of the strong field approximation used. The hexacarbonyls of chromium, molybedenum and tungsten all have ionisation potentials of around 8·4 eV, corresponding to loss of an electron from the $(t_{2g})^6$ closed shell [8]. The spectrum of $Fe(CO)_5$ [317] shows two bands of equal intensity in the d-electron ionisation region. These occur at binding energies of 8·6 and 9·9 eV, and may be assigned on simple molecular orbital arguments to the $d(e')$ and $d(e'')$ subshells respectively. Nickel carbonyl is of some interest, as calculations [318, 319] had disagreed on the ordering of the $d(t_2)$ and $d(e)$ subshells. The spectrum shows bands of binding energy 8·8 and 9·7, with areas in the ratio 3:2 respectively [8]; thus the t_2 level lies highest. The energies observed throughout appear consistent with a net positive charge on the metal.

The spectra of substituted carbonyls are more difficult to interpret. Substitution will lower the symmetry, causing further splitting of the d subshell. This could be of major interest in testing theories of bonding, despite the fact that the bands are close together and sometimes imperfectly resolved. Difficulties of interpretation could, however, arise from the electron redistribution energy discussed above. This could well be different for different orbitals, if only because the "positive hole" created by ionisation would be at different distances from the various ligands. Experimental problems arise from the relative involatility of many substituted carbonyls, especially since more volatile impurities or decomposition products could give rise to significant spurious peaks. Despite all these problems, a useful start has been made in collecting and interpreting the spectra of substituted carbonyls. For example, it has been shown that replacement of CO by PF_3 in a variety of carbonyl complexes increases the ionisation potentials, in accord with the view that PF_3 is a greater net acceptor of charge than is CO[320]. The spectra of a range of manganese derivatives has been reported. For the manganese pentacarbonyl halides, the first ionisation corresponds to removal of an electron from the $p(\pi)$ orbitals of the halogen, which are presumably raised in energy by interaction with the $d(xz, yz)$ orbitals of the metal (the halogen is regarded as lying on the z-axis). The latter orbitals are accordingly lower in energy (i.e. more bonding) than is the $d(xy)$ orbital. In the absence of the special effects of halide π-donation, the $d(xy)$ orbital would be expected to be more stabilised than $d(xz, yz)$, since it overlaps four rather than three carbonyl groups.

a)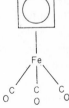

b)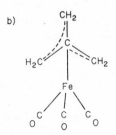

This is what is found in $Mn(CO)_5H$ and $Mn(CO)_5CF_3$, but in $Mn(CO)_5CH_3$ the more intense peak corresponds to the higher binding energy [321]. The reasons for this rather surprising result are not clear; but one suggestion is that the peak at higher binding energy does indeed correspond to $d(xz, yz)$, while the intensity of the second peak is due to the overlap of two ionisation bands, from $d(xy)$ and from CH_3-Mn σ orbitals respectively, In $Mn_2(CO)_{10}$, three bands are observed in the d-electron region, in the approximate intensity ratio $1:2:4$. The first ionisiation is thought to be from the electrons of the manganese-manganese σ-bond, while the second and third correspond to ionisations from various imperfectly resolved combinat-nations of the "t_{2g}" orbitals of the two metal atoms.

The low-energy photoelectron spectra of cyclobutadienetricarbonyliron (A) and trimethylenemethanetricarbonyliron (B) have been obtained, as have those of a range of iron tricarbonyl complexes of normal dienes, and the results interpreted in the light of calculations on the free ligands. In all cases the first ionisation corresponds to removal of an electron from a metal orbital, while the second corresponds essentially to ionisation of the highest ligand orbital, which is stabilised by interaction with the metal [322, 323].

Further developments in carbonyl low-energy photoelectron spectra may be expected to follow from the increasing availability of instruments with heated inlet systems. The problem of impurity peaks may be reduced as more spectra, including those due to possible volatile impurities, are accumulated. Areas of probable rapid expansion include the study of the effects of coordination to metal carbonyl groupings on the properties of unsaturated organic groups; the effects of substituents such as phosphines and amines on the energies of the metal electrons; and the valence shell structure of metal carbonyl cluster compounds.

High Energy Photoelectron Spectra of Carbonyls

The majority of high-energy investigations of metal carbonyl complexes have been concerned with the carbon and oxygen $1s$ levels. It is found that in the carbonyls of chromium, tungsten, iron and nickel the binding energy of carbon $1s$ is less than that of carbon $1s$ in free CO by amounts varying from $2·2$ to $2·5$ eV, while the binding energy of oxygen $1s$ is reduced by between $1·4$ and $1·7$ eV. These results confirm that coordinated CO is a net charge acceptor [324]. It is not surprising then that the binding energies of the metal electrons in $Cr(CO)_6$ take values typical of Cr(III) [325], while the core electrons of $Fe(CO)_5$ are as tightly held as those of iron (II) in $K_4Fe(CN)_6$ [326]. It is interesting that in $(\pi-C_5H_5) Mn(CO)_3$, the carbonyl carbon $1s$ binding energy is abnormally low, indicating that the CO group has accepted an unusual amount of charge. The ring carbons are even more positive than in other cyclopentadienyl complexes of metals, and the inference is that there has been a flow of charge from the ring to the CO groups. This finding may be related to the low value of the CO stretching frequencies in this complex compared with those of the unsubstituted carbonyls (see Chapter 7), and it would be of interest to know whether there is any general correlation between carbon $1s$ energies and CO stretching parameters.

One curious result has been claimed for some chromium carbonyl complexes. Satellite bands are observed corresponding to apparent carbon $1s$ binding energies around 6 eV higher than those responsible for the main band. The only possible interpretation for such a result is that on some occasions photoionisation is accompanied by another process, which absorbs around 6 eV of energy. It has been suggested that this process is formation of the cation in an excited state. Normally photoionisation is a pure one-electron process, but in this case it appears that electrons other than the one removed are being excited. A possible explanation is configuration interaction. The suggestion [327] is that the ground state of the complex is not very well represented by one configuration only. The relative weights of the different configurations change on ionisation, and this means that there is a chance that the ion generated will find itself in an excited state configuration.

8.2. Electronic Spectra of Metal Carbonyls

The electronic spectra of transition metal complexes, including carbonyls, can contain bands due to a variety of processes. These include internal ligand transitions, internal metal $(d–d)$ transitions, charge transfer from ligand to metal or from metal to ligand and, for polynuclear species, the excitation of electrons in the metal–metal bonding system.

Internal ligand transitions can be isolated for discussion when they involve transfer of electrons between levels that are relatively unaffected by complex formation. An example is the $\pi \to \pi^*$ spectrum of bipyridyl, which is scarcely altered in position or even in vibrational structure in the complexes (bipyridyl) $M(CO)_4$ (M = Cr, Mo, W) [328]. Where the levels involved in a transition are considerably modified by coordination, the transition acquires some charge transfer character and may be changed in position, intensity and general form. Some such process is presumably responsible for the absence of the characteristic structured UV bands of arenes from their tricarbonylchromium and related complexes. Internal ligand transitions due to CO are not expected at energies less than the vacuum UV. The lowest spin-allowed $\pi \to \pi^*$ and $n \to \pi^*$ transitions of free CO occur at 87,000 and 65,000 cm^{-1}, and both of these will be raised in energy on complex formation, as well as being split into several transitions differing in symmetry, intensity and energy.

All transition metal carbonyl complexes known or expected contain electrons assignable to the outermost d subshell of the metal atom; indeed, the availability of such electrons for π-back-donation appears to play an essential role in stabilising such carbonyls as isolable entities. With the exception of d^{10} complexes, such as those of Cu(I), metal carbonyls are therefore expected to show d–d transitions, though for reasons that will become apparent these may be obscured by bands of other kinds.

Even in a free gaseous atom or ion, the energy of a dn configuration is split into several "term" levels when n takes any value from 2 up to 8. This splitting arises because it is possible to arrange the d electrons in different ways, with different degrees of electron-electron repulsion. In such complexes as metal carbonyls, where the bonding interaction between metal and ligand strongly perturbs the d levels, the role of such term-term splitting is secondary to that of the splitting due to the ligand field. Those orbitals that point most directly towards the ligands are raised in energy relative to the others, and in carbonyl complexes the splitting so produced is modified, and commonly reinforced, by differences in stabilisation through π-bonding. In octahedral (or tetrahedral) complexes, the ligand field splitting may be represented by a single parameter, known as Δ. The energy required to promote an electron is closely related to this parameter, although in general not identical with it. This is because of changes in electron-electron repulsion energy on promotion. The mutual repulsion energies of electrons in pure d-orbitals may be expressed in terms of a very few parameters, two being sufficient for the changes caused by $d \to d$ transitions. In general, the promotion of a d-electron from one level to another gives rise to several bands, associated with different changes in electron-electron repulsion energy. If we impose on the d-orbitals of complexes the theory appropriate to free gaseous atoms and ions, it is possible in favourable cases to calculate values for the electron-electron repulsion parameters in the complex. These are always lowered relative to the free atom or ion, and the lowering (known as the nephelauxetic effect) is generally regarded as a measure of covalency.

One further feature of $d \to d$ transitions that should be mentioned is their lack of intensity. There are selection rules for electronic absorption, closely related to the corresponding rules for vibrational (infrared) absorption. In particular, transitions between orbitals of the same parity (odd or even behaviour under inversion) are electric dipole forbidden. Despite this, in complexes, such transitions invariably give rise to absorption. Two closely related mechanisms are involved. If the complex lacks a centre of symmetry, the so-called d-orbitals will mix with metal orbitals or with ligand orbital combinations of odd parity, giving the transition some odd $\to$ even character. The transition may thus become allowed, provided the product of the representation of the level from which the electron is excited to that of the representation to which it is promoted includes the representation of x-, y-, or z-axis orientated electric dipoles. Even if the complex has a centre of symmetry, or if the transition of interest is forbidden in the actual symmetry, absorption can still take place by what is known as the vibronic mechanism. This involves a combination of vibrational and electronic excitations, so that the resultant overall excitation is symmetry allowed. Generally, vibronic excitations are weaker than those $d \rightarrow d$ excitations that become allowed through the influence of the static environment, and these in turn are weaker than the fully allowed bands discussed below. Another identifying feature of vibronic transitions is that they lose some (but not all) intensity on cooling, since transitions from vibrationally excited levels are more allowed than the corresponding transitions from the vibrational groundstate. In carbonyl complexes, the intensity of d–d transitions is expected to be relatively high, with extinction coefficients of the order of one thousand, since covalent overlap leads to efficient perturbation of the form of the d-orbitals by static or vibronic distortions.

Charge transfer excitations involve the transfer of an electron from an orbital located primarily on one atom (or set of atoms) to an orbital located primarily on others. Both metal to ligand and ligand to metal transitions are possible. The latter, however, are expected to lie in metal carbonyls at the high energy end of the quartz ultraviolet region, if the 18-electron rule is obeyed. Where fewer than 18 electrons are associated with the metal, as in complexes of V(O) and Cr(I), low energy transitions occur to vacant σ-non-bonding levels of the metal, presumably from the π-bonding levels of CO. Charge transfer from the metal to the formally vacant 2π levels of co-ordinated CO can take place in the near ultraviolet. This charge transfer is potentially a highly informative process, since the CO 2π generally give rise in complexes to a number of levels with different degrees of π-interaction.

The interpretation of electronic spectra in general, and of charge transfer spectra in particular, is made more difficult by the high energy of the excited states, and by the two-orbital nature of the electronic transitions. The high energy of an excited state configuration will inevitably bring it closer in energy to a large number of other excited configurations of the same symmetry. This situation is liable to lead to a marked degree of configuration interaction (some configuration interaction is present even in the ground state), reducing the energy of the lowest excited states by an amount that is difficult to estimate. The two-orbital nature of the transition means that the energy required will depend on the energy of the orbital from which an electron is excited, the energy of the orbital to which it is promoted, and the interaction between them. This interaction can be thought of as an electrostatic attraction between the excited electron and the positive hole left in the ground state configuration by its promotion.

For these reasons it is not legitimate to equate the energy of an electronic transition with the energy difference of the orbitals between which the electron jumps. Quite apart from the effects of electronic reorganisation (including the changes in configuration interaction) the energy in the ground state of the upper orbital represents the electron affinity of the upper orbital when the lower orbitals are full. The energy of the same orbital in the excited state, however, represents the electron affinity of this orbital when one electron is missing from a lower orbital. This could explain the fact that although a large number of calculations place the $d(e_g)$ levels of the metal hexacarbonyls at far higher energy than the lowest empty ligand orbitals, the lowest energy bands in the spectrum are due to $d \rightarrow d$ transitions. A more trivial consequence of the two-electron nature of electronic processes is the existence of a large number of separate transitions, especially in environments of low symmetry. This causes obvious difficulties of observation and assignment.

In view of the complex nature of electronic transitions, it is not surprising that the only carbonyls for which the spectra are well understood are the hexacarbonyls of Group VI and related species [7, 27].† In the ground state, the highest occupied electronic energy level is the t_{2g} set that corresponds approximately to $d(xy, yz, zx)$, and the ground term is $^1A_{1g}(t_{2g}^6)$. Promotion of an electron within the d subshell gives rise to terms $^1T_{1g}$, $^1T_{2g}$, $^3T_{1g}$, $^3T_{2g}$, all derived from the configuration $t_{2g}^5 e_g$. These terms differ in energy because of different degrees of mutual repulsion between the d-electrons. Four d–d transitions may therefore be expected in the electronic spectrum. Of these, two will be moderately weak, since they

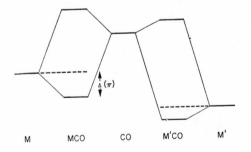

M   MCO   CO   M'CO   M'

FIG. 8.2. Interaction between $d(\pi)$ orbital on Metal and π^* Orbital of CO Stabilising $d(\pi)$ and Increasing d–d Excitation Energies by $\Delta(\pi)$. Note Larger Effect for M than for More Electron-attracting M'.

† For a dissenting interpretation even in this case see [331].

violate the selection rule that requires a change in parity on excitation. The remaining two, since they give rise to triplet states of even parity, are spin- as well as parity-forbidden, and exceedingly weak. The $d \to d$ transitions have $n \to \sigma^*$ (or indeed, in these complexes, $\pi \to \sigma^*$) character, and therefore give rise to a considerable change in equilibrium distances. A consequence (see Fig. 8.1) is that the bands are broad, and in some cases vibrational structure can actually be observed at low temperature.

TABLE 8.1. Ligand field parameters (cm^{-1}) for some hexacarbonyls (a).

Species	Δ	B' (b)	B (c)	C'	$\beta = B'/B$
$V(CO)_6^-$	25,500	430	790	1300 est	
$Cr(CO)_6$	32,200	520	790	1700 est	0·66
$Mn(CO)_6^+$	41,650	—	870	2600	
$Mo(CO)_6$	32,150	350	460	1100	0·83
$W(CO)_6$	32,200	390	370	1300	1·1
$Re(CO)_6^+$	41,000	470	470	1375	1·0

(a) Data from [7].
(b) Apparent value of parameter B in complex.
(c) Free atom (or ion) value.

The frequencies of the $d \to d$ transitions, and the values of the ligand field splitting parameter (Δ) and of the electron-electron repulsion parameters (B and C) estimated from these, are included in Table (8.1). The value of Δ is high, and depends relatively little on charge. This is as expected from the important role assigned to π-back-donation of the t_{2g} electrons, since this back-bonding will most effectively stabilise the t_{2g} orbitals, and increase Δ, where the formal oxidation number of the metal is low (Fig. 8.2).

The most prominent feature of the spectra of the hexacarbonyls is a pair of very strong peaks in the ultraviolet, with a high energy shoulder to the higher energy peak. These bands are all too intense to be assigned to $d \to d$ transitions, and all move to higher energy on going from $V(CO)_6^-$ to $Cr(CO)_6$, and from $W(CO)_6$ to $Re(CO)_6^+$. They are accordingly assigned to charge transfer from the metal t_{2g} orbitals to various combinations of CO 2π. The latter combine to span the irreducible representations $t_{1g} + t_{1u} + t_{2g} + t_{2u}$ of O_h. The t_{1u} orbital is then capable of being lowered in energy by interaction with the vacant metal $(n + 1)p$ orbital, while the $t_{2g}(CO2\pi)$ orbital must be raised by interaction with t_{2g} (metal d). The lower energy of the two intense bands has on these grounds been assigned to a transition $t_{2g}(d) \to t_{1u}(2\pi)$, while the higher energy band is assigned as $t_{2g}(d) \to t_{2u}(2\pi)$.

The higher energy shoulder of the latter band may tentatively be assigned to a $t_{2g}(d) \rightarrow t_{2g}(2\pi)$ charge transfer, which is formally forbidden and therefore expected to give a weakened charge transfer band [329]. As required by this assignment, the separation between the main band and the shoulder decreases with increasing net positive charge on the metal [7, 27].

The ultraviolet spectra of other carbonyl complexes have been comparatively little studied. Iron pentacarbonyl shows maxima at 35,000 and 41,500 cm^{-1}, assignable to the D_{3h}-allowed $d \rightarrow d$ transition $e'(xy, x^2 - y^2) \rightarrow a_1'(z^2)$ and to a particular set of metal to ligand charge transfer bands respectively, and a broad featureless absorption which is still increasing at 50,000 cm^{-1} [330]; the $d \rightarrow d$ transition $e''(xy, yz) \rightarrow a_1'(z^2)$ is forbidden in D_{3h}. The spectrum of nickel tetracarbonyl is, of course, devoid of $d \rightarrow d$ bands and consists of a continuous absorption increasing throughout the ultraviolet to a maximum at 49,600 cm^{-1}, with shoulders at 44,600 and 42,300 cm^{-1} [331]. The spectra of manganese pentacarbonyl chloride and bromide have been recorded. These spectra contain a moderately strong band at around 26,000 cm^{-1}, which was assigned to charge transfer from metal to the 2π orbital of the axial CO group (i.e. that trans to the halogen) [332]. Monosubstituted derivatives of the Group VI hexacarbonyls also show a band of moderate intensity in the near ultraviolet. It has been shown that, at least where the substituent is a phosphine, this band does not lose intensity down to $-185°C$, strongly suggesting that it is due to an electric dipole allowed transition. Moreover, the band lies between a similar band of comparable intensity in the *trans*-disubstituted species, and the lowest charge transfer band of the hexacarbonyls. Thus the relevant transition would appear to be charge transfer from the $d(xy, yz)$ orbitals of the metal to a combination of *equatorial* CO 2π orbitals, of z-like symmetry [333, 85]. This result also explains the dependence of the frequency of the band in monosubstituted species on the nature of the substituent ligand. Good π-acceptors stabilise the donor d-orbitals, thus raising the energy of the transition. In the absence of π-bonding effects, σ-donors may perhaps slightly raise the required energy by direct interaction with the relevant acceptor orbital. Such direct interaction between an axial ligand and equatorial carbonyl groups has independently been suggested on the basis of *a priori* calculations [334, 335].

It is difficult to say much of value about the electronic spectra of more heavily substituted carbonyls. Metal to carbonyl charge transfer bands are presumably to be expected throughout the ultraviolet region, but the ordering and nature of the bands is difficult to predict. One particularly elegant study of a metal to ligand charge transfer band deserves mention, although strictly it lies outside the scope of this chapter, because of the way in which the results were related to infrared data. Substituted carbonyls of the type (*C*) (M = Cr, Mo, W) show strong characteristic bands at around 25,000 cm^{-1} which are assigned to charge transfer from the metal to the vacant π^* orbitals of the nitrogen ligand. These bands show a marked solvatochromism, being raised in energy by more polar solvents. The effect is reduced on replacing one CO group with triphenylphosphine, and can be reversed by further substitution. Since polar solvents will stabilise the ground state dipoles, the inference is that in *C* the charge transfer reduces or reverses this dipole, which therefore must be in the sense (ligand positive, metal negative). Replacing CO by triphenylphosphine reduces the electron demand of the metal, so that the highest filled orbital becomes more concentrated on the ligand. Thus in the heavily substituted species, the transition shifts charge from ligand to metal, in the same sense as the existing dipole. An important peripheral result is that the modes due to vibrations of the CO groups *trans* to nitrogen can be distinguished from the other modes by their greater solvent dependence. A polar solvent should increase the ground state donation from diazabutadiene to metal, thereby lowering the frequencies, as observed [336].

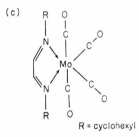

R = cyclohexyl

8.3. Carbon-13 NMR Spectra

The use of ^{13}C n.m.r. spectroscopy has been hampered by the low natural abundance of the nucleus of interest and, in the case of metal carbonyls, its long relaxation time. This has meant that samples of metal carbonyls rapidly become "saturated" in the spectrometer beam, with equal populations in the ground and excited states and no net absorption.

The problem of signal weakness due to low natural abundance has been greatly reduced by advances in instrumentation and in particular the development of n.m.r. instruments that operate on the Fourier principle. The problem of long relaxation times appears to have been solved very recently. Addition of a paramagnetic transition metal ion subjects the nuclei of interest to a fluctuating magnetic field and thereby accelerates relaxation. If the paramagnetic ion used forms a complex with the sample species, large shifts may be induced,† but it has now been found that in many cases Cr(III) induces relaxation without causing a shift [337]. Coupling to ligand protons introduces unwanted complexity, and is conventionally removed by deliberate saturation.

TABLE 8.2. Some ^{13}C NMR Data for Metal Carbonyls (a).

Species	(b)	$^2J(^{31}P^{13}C)(c)$	$^1J(M^{13}C)$	
$Cr(CO)_6$	212·1			
$Mo(CO)_6$	202·0		68	
$W(CO)_6$	192·1		125·0	
$W(CO)_5P(Me)_3$	199·5	37·2	139·1	(axial)
	196·5	11·6	125·1	(radial)
$W(CO)_5PBu''_3$	200·4	18·9	142·1	(axial)
	198·6	7·3	124·4	(radial)
$W(CO)_5AsPh_3$	199·7			(axial)
	197'5		124	(radial)
$W(CO)_5NH_2C_6H_{11}$	201·9			(axial)
	199·1		132	(radial)
$W(CO)_5C(OMe)Ph$	204·6			(axial)
	198·6		125·0	(radial)
$W(CO)_3(durene)$ (d)	213·7		189	

(a) Data from [338] and refs. therein.
(b) p.p.m. position downfield from internal Me$_4$Si.
(c) Hz.
(d) Data from [340].

There are two main quantities of chemical interest obtainable from an n.m.r. spectrum. These are chemical shifts and coupling constants. Additional kinetic information is obtained in some cases. For instance, if two chemically distinct nuclei give a single combined signal, it is probable that they are exchanging environments at a rate greater than the difference between their resonance frequencies. The precise origin of the chemical shifts in coordinated CO is still a matter of controversy [338, and references therein]. Possible factors are inductive effects, π-electron densities, and the size of the energy separation between the ground state and those excited states that affect the carbon 2p orbitals. It is found experimentally (see Table 8.2) that replacement of CO by a donor L in the

† This effect is itself a possible source of useful information about the donor power of carbonyl complexes, their stereochemistry, and the identity of the donor site. Much of this information is already available from ^{1}H n.m.r. studies [50], although there are some species of great interest (such as $V(CO)_6^-$ and $Co(CO)_4^-$) that contain no protons.

Group VI hexacarbonyls increases δ, the downfield chemical shift (ppm relative to tetramethylsilane), of coordinated CO. This shift is greater for the CO group *trans* to the substituent, and increases along the series

$$CO < phosphite < phosphine, arsine < amine < carbene$$

i.e. with increasing ligand σ-donor power. Moreover, the observed shift of carbonyl *cis* to a substituent is greater for *trans* disubstituted species than it is for the corresponding monosubstututed complexes, and for a given Group VI metal δ corelates with the lowering both of CO stretching frequency and of the energy of the band assigned in the previous section to the lowest metal $d(xz, yz) \rightarrow CO(axial) 2\pi$ charge transfer process. The chemical shift of carbon is thought to be dominated by a paramagnetic shielding term, which may be represented as

$$\sigma_p = K \langle r^{-3} \rangle_{2p} (Q_{AA} + Q_{AB})/\Delta E \tag{8.1}$$

Here K is a constant, ΔE is an average excitation energy, and $\langle r^{-3} \rangle_{2p}$ is the average expectation value of r^{-3} for the $2p$ orbitals. Q_{AA} depends on the local charge density and is essentially constant for most carbon nuclei. Q_{AB} arises because of the effects of the neighbouring atom B in inducing current on atom A, and requires σ- and π-bonding between A and B. It is thus easy to see how the chemical shift can be affected by ΔE, by inductive effects on $\langle r^{-3} \rangle$, and directly by metal–carbon π-bonding, but the relative importance of these various effects is still a matter for argument.

Coupling may be observed between carbonyl carbon and a suitable metal nucleus, and also between carbonyl carbon and nuclei in other ligands. The general trends in coupling between carbonyl carbon and the phosphorus of phosphine and phosphite ligands resemble trends in phosphorus-phosphorus coupling, and may thus be ascribed to changes in s-electron densities in the bonds [339]. Coupling between tungsten and equatorial carbon in complexes of the type $W(CO)_5L$ and *trans*-$W(CO)_4L_2$ is little affected by the nature of L; but coupling to the axial carbonyl carbon increases from $W(CO)_6$ through $W(CO)_5$(phosphite) to $W(CO)_5$(phosphine), and is considerably greater for (durene)$W(CO)_3$ [338, 340].

The spectra of $Fe(CO)_5$ and some substituted derivatives have been reported [341]. There is no splitting of the single signal due to carbonyl carbon, or of the satellite bands due to coupling between carbon and ^{57}Fe, so that rapid intramolecular averaging of axial and equatorial carbon must be taking place. Early attempts to observe the ^{13}C spectra of $Mn_2(CO)_{10}$ and $Co_2(CO)_8$ were unsuccessful [342], possibly because of quadrupole relaxation by ^{55}Mn and ^{59}Co. A single signal was detected for ^{17}O in $Mn_2(CO)_{10}$, although rapid exchange of axial and equatorial CO seems unlikely. It is known that $Co_2(CO)_8$ exists in solution as an equilibrium mixture of bridged and unbridged isomers, but nothing at all is known about the rate or pathway of isomersation. The observation of the ^{13}CO spectra of these carbonyls would, therefore, be of great interest.

Degenerate isomerisation has been observed [369, 370] in ^{13}CO-enriched $Rh_4(CO)_{12}$ (isostructural [371] with $Co_4(CO)_{12}$); the expected complex low-temperature ^{13}C n.m.r. spectrum collapses uniformly to give a 1:4:6:4:1 quintet, as required for coupling of equivalent carbons to four rhodium atoms (^{103}Rh 100%; $I = \frac{1}{2}$). The low-temperature spectrum of $RhCo_3(CO)_{12}$ establishes a structure related to that of $Co_4(CO)_{12}$ by the replacement of one basal cobalt with rhodium. In this case, averaging is clearly non-uniform. Between the low-temperature limit and 10°C, the two terminal carbonyls on rhodium remain aloof from averaging processes affecting all the other CO groups, and enter into averaging only between 10°C and 30°C [372]. The important inference is that the uniform averaging in $Rh_4(CO)_{12}$ need not necessarily take place *via* symmetric intermediates, provided the three equivalent sides of the Rh(apical) [Rh(basal)]$_4$ pyramid can each take part in some concerted averaging processes [373].

Glossary of Symbols

Standard chemical and group-theoretical symbols are omitted, as are various symbols that occur at one point only in the Text. Symbols used to specify the energy-factored force fields of metal carbonyls are collected separately on pp. 42, 43.

A	Amplitude of vibration; optical absorbance
$\mathbf{C}$	Compliance constant matrix
D	Optical density
$\mathbf{E}$	Unit matrix
E	Energy
$\mathbf{F}$	Force constant matrix, where elements F_{ii}, F_{ij} are the the force constants and interaction constants of the Generalized Valence Force Field.
f_{ij}	Coefficient of mass-weighted coordinate product $q_i\,q_j$ in potential energy expression (Eq. 2.15).
$\mathbf{G}$	Effective reciprocal mass matrix.
G_{ii}	Anharmonicity coefficient of l_i for ith (degenerate) mode.
g_i	Degeneracy of ith mode.
h	Planck's Constant
$I(I_0)$	Intensity of transmitted (incident) light
I	Intensity, integrated, of absorption band
I_x	Moment of inertia around x-axis
i	Interaction parameter
K	Effective CO parameter for normal mode
k	Force constant; stretching parameter
$\bar{k}$	Mean stretching parameter
$k(m)$	Effective stretching parameter for m'th CO symmetry mode

$k(\text{int})$	Effective interaction parameter
L_{ij}	Element of matrix linking symmetry coordinate S_i with normal coordinate Q_j (Eq. 2.41)
l	Optical pathlength
l_i	Internal rotational quantum number (p.9) if ith degenerate mode.
l_{ij}	Coefficient linking bond distortion r_i with symmetry coordinate S_j (Eq. 3.35)
m	Order of F-matrix
m_i	Mass of ith nucleus
n	Number of nuclei in molecule
Q	Normal coordinate
q	Displacement
q_i	Mass-weighted coordinate
$\mathbf{R}; \overline{\mathbf{R}}; \overline{R}$	Raman scattering tensor; average Raman scattering tensor; mean value of Raman scattering tensor.
$R(^{13}CO), R(C^{18}O)$	Pracitical values for $v(^{13}C^{16}O)/v(^{12}C^{16}O)$, $v(^{12}C^{18}O)/v(^{12}C^{16}O)$
r	General coordinate; coordinate for distortion of individual CO group
$S(v)$	Slit function
S	Symmetry coordinate
s	Internal coordinate
T	Transmittance (commonly, percentage transmittance) of monochromatic light
T	Kinetic energy
t_{ij}	General kinetic energy coefficient (Eq. 2.27)
u_j	Cartesian coordinate
V	Potential energy
V_0	Potential energy at equilibrium
v	Vibrational quantum number
v_{ij}	General potential energy coefficient (Eq. 2.28)
$v_i^{xy}(v_i^{xy,\,zw})$	Potential energy due to distortion of bond XY, (to interaction of distortion of bonds XY, ZW) during ith vibration
X_{kl}, X'_{ki}	Terms in vibrational energy of anharmonic molecule, Eqs. (2.11, 2.12).
x_i, y_i, z_i	Cartesian coordinates of ith particle

α	Polarisability tensor; angle
β	angle
$\gamma(R)$	Anisotropy of Raman scattering tensor $\mathbf{R}$
$\Delta(j)$	Distortion of the jth MCO group
δ	Perturbation of CO parameter k
$\delta(XY)$	Change in XY bandlength
δ_{ij}	Kroneker delta function, unity for $i = j$, otherwise zero
ε	Molar extinction coefficient
θ; $\theta(i)$	Angle; angle between MCO groups (assumed linear) connected by interaction parameter i
λ_k	kth eigenvalue of force constant matrix, corresponding to normal coordinate Q_k (Eqs. 2.20–2.26).
μ	Reduced mass, especially of CO group; dipole moment
μ^*	Reduced mass of isotopically labelled species (especially of $^{13}C^{16}O$, $^{12}C^{18}O$)
$\mu(j)$	Dipole due to unit distortion of jth CO group
ν	Frequency, especially of a photon or of a molecular vibration (Hz); also, loosely, wavenumber (cm^{-1}).
ν_0, ν'	Frequency of incident, of scattered light
ρ_l, ρ_n	Depolarisation ratio for plane polarisation light, for natural light.
ϕ	Phase; angle or angular parameter
ω	Harmonic frequency
ω_k'	Term in energy expression for anharmonic molecule (Eq. 2.12)
ω^*	Harmonic frequency of isotopically substituted molecule

Appendix II

F and G Matrix Elements

We give here expressions for **G**-matrix elements connecting the internal coordinates of species $M(CO)_x L_y$. The MCO groups are assumed linear, and the L groups are treated as single atoms, ignoring coupling between the skeletal motions of L and the rest of the molecule. The treatment of such coupling requires more general expressions, which have been given [13, Appendix VI, and references therein]. The **G**-matrix elements are readily re-written in terms of the elements for the internal coordinates, and the procedure is illustrated by reference to the model system $M(CO)_2$.

We use the symbols $g(A)$, for the reciprocal mass of atom A, and $G(AB, AB)$, $G(AB, ABC)$ etc. for the **G**-matrix elements connecting $v(AB)$ with itself, $v(AB)$ with $\delta(ABC)$, etc., while $\rho(AB)$ is the reciprocal of $r(AB)$. The **G**-matrix elements linking distortions with no atom in common are zero.

The matrix elements within individual MCO or ML fragments are

$$G(MC) \quad = g(M) + g(C)$$

$$G(CO) \quad = g(C) + g(O)$$

$$G(MC, CO) \quad = -g(C)$$

$$G(MCO) \quad = \rho(MC)^2\, g(M) + \rho(CO)^2\, g(O) + [\rho(MC) + \rho(CO)]^2\, g(C)$$

$$G(MC, MCO) = G(CO, MCO) = 0$$

$$G(ML) \quad = g(M) + g(L) \tag{II.1}$$

In writing down the elements linking different fragments, we distinguish here between different atoms of the same element. This is to allow for the possibilities of partial isotopic substitution and, more importantly, of differences in chemical environment and bondlength. The relevant elements may be classified as stretch–stretch, stretch–bend, and bend–bend interactions:

$$G(MC, MC') = g(M) \cos (\angle \, CMC')$$
$$G(MC, ML) = g(M) \cos (\angle \, CML)$$
$$G(ML, ML') = g(M) \cos (\angle \, LML') \tag{II.2}$$
$$G(MCO, MC') = \rho(MC) \, g(M) \sin (\angle \, CMC') \text{ for the}$$
$$\text{MCO bend in the CMC' plane}$$

$$= 0 \text{ for the MCO bend perpendicular to the CMC' plane.}$$

Similarly,

$$G(MCO, ML) = \rho(MC) \, g(M) \sin (\angle \, CML) \text{ or } = 0 \tag{II.3}$$

$$G(CMC', MC) = - \, \rho(MC') \, g(M) \sin (\angle \, CMC') \tag{II.4}$$

$$G(CMC', ML) = - \, [\rho(MC) \sin (\angle \, LMC) \cos \psi \, (LCC')$$
$$+ \rho(MC') \sin (\angle \, LMC') \cos \psi \, (LC'C)] \, g(M) \tag{II.5}$$

where ψ (LCC') is the dihedral angle defined by

$$\cos \psi \, (LCC') = [\cos (\angle \, LMC') - \cos (\angle \, LMC) \cos (\angle \, CMC')]$$
$$\div \sin (\angle \, LMC) \sin (\angle \, CMC') \tag{II.6}$$

$$G(CML, MC) = - \, \rho(ML) \, g(M) \sin (\angle \, LMC)$$

$$G(CML, ML) = - \, \rho(MC) \, g(M) \sin (\angle \, CML)$$

$$G(CML, MC') = - \, [\rho(ML) \sin (\angle \, C'ML) \cos \psi \, (C'LC)$$
$$+ \rho(MC) \sin (\angle \, C'MC) \cos \psi \, (C'CL)] \, g(M)$$

$$G(CML, ML') = - \, [\rho(ML) \sin \angle \, (L'ML) \cos \psi \, (L'LC)$$
$$+ \rho(MC) \sin (L'MC) \cos \psi \, (L'CL)] \, g(M) \tag{II.7}$$

$$G(LML', ML) = - \, \rho(ML') \, g(M) \sin (\angle \, L'ML)$$

$$G(LML', MC) = - \, [\rho(ML) \sin (\angle \, CML) \cos \psi \, (CLL')$$
$$+ \rho(ML') \sin (\angle \, CML') \cos \psi \, (CL'L)] g(M) \tag{II.8}$$

$$G(MCO, MC'O') = \cos (\angle \, CMC') \rho(MC) \, \rho(MC') \, g(M) \text{ (both bends in-plane)}$$

$$= \rho(MC) \, \rho(MC') g(M) \text{ (both out-of-plane)}$$

$$= 0 \text{ (one in-plane, one out-of-plane)} \tag{II.9}$$

$$G(\text{MCO, CMC}') = -\rho(\text{MC})\{[\rho(\text{MC}) - \rho(\text{MC}')\cos(\angle\,\text{CMC}')]g(\text{M})$$
$$+ [\rho(\text{MC}) + \rho(\text{CO})]g(\text{C})\} \text{ for in-phase MCO bend}$$
$$= 0 \text{ for out-of-phase MCO bend} \qquad (\text{II.10})$$

$G(\text{MCO, CML})$ is similar in form to $G(\text{MCO, CMC}')$

$$G(\text{MCO, LML}') = \{[\sin(\angle\,\text{CML}')\cos(\angle\,\text{LML}')\cos(\tau\text{L}')$$
$$- \sin(\angle\,\text{CML})\cos(\tau\text{L})]\,\rho(\text{ML}')$$
$$+ [\sin(\angle\,\text{CML})\cos(\angle\,\text{LML}')\cos(\tau\text{L})$$
$$- \sin(\angle\,\text{CML}')\cos(\tau\text{L}')]\,\rho(\text{ML})\}$$
$$\times \rho(\text{MC})g(\text{M})/\sin(\angle\,\text{L}'\text{ML}') \qquad (\text{II.11})$$

where τL is the angle between the CML plane and the plane of the MCO bend.

$$G(\text{CMC}', \text{CML}) = \rho^2(\text{MC})\cos\psi\,(\text{C}'\text{CL})g(\text{C}) + \{[\rho(\text{CM}) - \rho(\text{MC}')\cos$$
$$(\angle\,\text{CMC}') - \rho(\text{ML})\cos(\angle\,\text{CML})]\,\rho(\text{CM})\cos\psi$$
$$(\text{C}'\text{CL}) + [\sin(\angle\,\text{CMC}')\sin(\angle\,\text{CML})\sin^2\psi\,(\text{C}'\text{CL})$$
$$+ \cos(\angle\,\text{C}'\text{ML})\cos\psi\,(\text{C}'\text{CL})]\,\rho(\text{MC}')\,\rho(\text{ML})\}g(\text{M})$$
$$(\text{II.12})$$

$$G(\text{CML, C}'\text{ML}') = \{[\cos(\angle\,\text{LML}') - \cos(\angle\,\text{C}'\text{ML})\cos(\angle\,\text{C}'\text{ML}')$$
$$- \cos(\angle\,\text{CML})\cos(\angle\,\text{CML}') + \cos(\angle\,\text{CMC}')$$
$$\cos(\angle\,\text{CML})\cos(\angle\,\text{C}'\text{ML}')]\,\rho(\text{MC})\,\rho(\text{MC}')$$
$$+ [\cos\text{LMC}') - \cos(\angle\,\text{L}'\text{ML})\cos(\angle\,\text{L}'\text{MC}')$$
$$- \cos(\angle\,\text{CML})\cos(\angle\,\text{CMC}') + \cos(\angle\,\text{CML}')$$
$$\cos(\angle\,\text{CML})\cos(\angle\,\text{L}'\text{MC}')]\,\rho(\text{MC})\,\rho(\text{ML}')$$
$$+ [\cos(\angle\,\text{CML}') - \cos(\angle\,\text{C}'\text{MC})\cos(\angle\,\text{C}'\text{ML}')$$
$$- \cos(\angle\,\text{LMC})\cos(\angle\,\text{LML}') + \cos(\angle\,\text{LMC})$$
$$\cos(\angle\,\text{LMC}')\cos(\angle\,\text{C}'\text{ML}')]\,\rho(\text{ML})\,\rho(\text{MC}')$$
$$+ [\cos(\angle\,\text{CMC}') - \cos(\angle\,\text{L}'\text{MC})\cos(\angle\,\text{L}'\text{MC}')$$
$$- \cos(\angle\,\text{LMC})\cos(\angle\,\text{LMC}')$$
$$+ \cos(\angle\,\text{LML}')\cos(\angle\,\text{LMC})\cos(\angle\,\text{L}'\text{MC}')]$$
$$\times \rho(\text{ML})\,\rho(\text{ML}')\} \times g(\text{M})/\sin(\angle\,\text{CML})\sin(\angle\,\text{C}'\text{ML}')$$
$$(\text{II.13})$$

$G(CMC', LML')$ may be found by interchanging C' and L in Eq. (II.13). The expression for $G(CML, CML')$ may be derived by writing L, L' for C', L in Eq. (II.12); $G(LMC, LML')$ may be found similarly.

The G-matrix elements linking symmetry coordinates may be found very simply from the elements linking internal coordinates. For example, for the symmetric and antisymmetric metal–carbon stretches of a fragment $M(CO)_2$ of C_s symmetry we have

$$G\{A'(MC), A'(MC)\} = G\{MC_1 + MC_2)/\sqrt{2}, (MC_1 + MC_2)/\sqrt{2}\}$$

$$= G(MC, MC) + G(MC, MC')$$

$$= g(M) + g(C) + g(M) \cos (\angle C_1 MC_2).$$

$$G\{A'(MC), A''(MC)\} = [G(MC_1, MC_1) - G(MC_2, MC_2)]/2 = 0$$

The forms of **F** matrices generally follow directly from the definitions of the force constants and interaction constants. The force field may be specified by reference to any set of parameters that fully define the internal coordinates of the atoms. The particular parameters used in this book, and in current discussions of the spectra of carbonyls, are the bond lengths and bond angles. Other possibly significant parameters, such as non-bonded distances, are not discussed as such.

The definitions of force constants and interaction constants involving stretches only are straightforward. The angle $C_i MC_j$ is known as α_{ij}. The MCO angles are known as β, and in an octahedral carbonyl the distortion of MC_iO_i in the C_iMC_j plane is referred to as a distortion of β_{ij}. In tetrahedral carbonyls, $\beta_{1a}, \beta_{2a}, \beta_{4a}$ are distortions of MC_1O_2 in the C_1MC_2 plane, and of MC_3O_3, MC_4O_4 in the C_3MC_4 plane, while β_{ib} is a distortion perpendicular to β_{ia}. In the notation of Fig. 3.21 for octahedral carbonyls, the stretch–bend and bend–bend interaction constants are

$F'_{MC,\alpha}$	MC_1 with α_{12}
$F'_{MC,\beta}$	MC_1 with β_{21}
$F_{\beta\beta}{}'$	β_{12} with β_{32}
$F_{\beta\beta}{}''$	β_{12} with β_{21}
$F_{\beta\beta}{}'''$	β_{12} with β_{52}
$F_{\alpha\alpha}{}'$	α_{12} with α_{15}
$F_{\alpha\alpha}{}''$	α_{12} with α_{34}
$F_{\alpha\alpha}{}'''$	α_{12} with α_{53}
$F_{\alpha\alpha}{}''''$	α_{12} with α_{15}

f_α coefficient of linear term, $\sum \alpha_{ij}$, in potential energy function

$\bar{F}_{\alpha\alpha}{}'$ $F_{\alpha\alpha}{}' - f_\alpha$

$F_{\alpha\beta}{}'$ α_{12} with β_{12}

$F_{\alpha\beta}{}''$ α_{12} with β_{32}

$F_{\alpha\beta}{}'''$ α_{12} with β_{52}

For tetrahedral carbonyls, the interaction constants are

$F'_{MC,CO}$	MC_i with C_iO_i
$F''_{MC,CO}$	MC_i with C_jO_j
$F''_{MC,MC}$	MC_i with C_jO_j
$F_{\beta\beta}{}'$	β_{1a} with β_{2a}
$F_{\beta\beta}{}''$	β_{1a} with β_{3b}
$F_{\beta\beta}{}'''$	β_{1a} with β_{3a}
$F_{\alpha\alpha}{}'$	α_{12} with α_{23}
$F_{\alpha\alpha}{}''$	α_{12} with α_{34}
$F_{\beta\alpha}{}'$	β_{1a} with α_{12}
$F_{\beta\alpha}{}''$	β_{1a} with α_{13}
$F_{\beta\alpha}{}'''$	β_{1a} with α_{23}
$F_{\beta\alpha}{}''''$	β_{1a} with α_{34}
$F_{r,\beta}{}'$	β_{1a} with r_2
$F_{r,\beta}{}''$	β_{1a} with r_3
$F_{r,\alpha}{}'$	α_{12} with r_1
$F_{r,a}{}''$	α_{12} with r_3

It is obvious that in substituted carbonyls the number of interaction constants will be large, and the effects of the proliferation of unknowns on the energy factored force field is discussed at several points throughout this book. Whether values for the other interaction constants can be transferred from one species to another is at present unknown.

The Symmetry Classification of Combination and Overtone Bands

The symmetry classification of combination bands, including combinations of distinct modes belonging to the same degenerate representation, may be found by direct multiplication (Eq. (6.1)).

If a vibrational mode belongs to a non-degenerate real representation of the molecular point-group, then its binary overtone belongs to the fully symmetric representation, while the ternary overtone belongs to the same representation as the fundamental.

The symmetry of a ternary combination-overtone excited state $(v_a = 2, v_b = 1)$ is given by the direct product of the symmetries of the binary overtone of v_a and the fundamental of v_b.

The symmetries of binary and ternary overtones of degenerate states follow from Eqs (6.2)–(6.5):

$$C_3 : v(E) = 2, \Gamma = E + A; v(E) = 3, \Gamma = E + 2A$$

$$\left.\begin{matrix} C_{3v} \\ D_3 \end{matrix}\right\} v(E) = 2, \Gamma = E + A_1; v(E) = 3, \Gamma = E + A_1 + A_2$$

$$C_4 : \quad v(E) = 2, \Gamma = E + A; v(E) = 3, \Gamma = E + 2A$$

$$\left.\begin{matrix} C_{4v} : \\ D_4 : \\ D_{2d} : \end{matrix}\right\} v(E) = 2, \Gamma = A_1 + B_1 + B_2; v(E) = 3, \Gamma = 2E$$

$$D_{3h} : \quad v(E') = 2, \Gamma = E' + A_1'; v(E') = 3, \Gamma = E' + A_1' + A_2'$$
$$v(E'') = 2, \Gamma = E' + A_1'; v(E'') = 3, \Gamma = E'' + A_1'' + A_2''$$

$T:$ $\quad v(E) = 2, \Gamma = E + A; v(E) = 3, \Gamma = E + 2A$

$\quad\quad v(T) = 2, \Gamma = T + E + A; v(T) = 3, \Gamma = 3T + A$

$T_d:$
$O.$
$\begin{cases} v(E) = 2, \Gamma = E + A_1; v(E) = 3, \Gamma = E + A_1 + A_2 \\ v(T_1) = 2, \Gamma = T_2 + E + A_1; v(T_1) = 3, \Gamma = 2T_1 + T_2 + A_2 \\ v(T_2) = 2, \Gamma = T_2 + E + A_1; v(T_2) = 3, \Gamma = T_1 + 2T_2 + A_1 \end{cases}$

$D_{3d} = D_3 \times C_i, D_{4h} = D_4 \times C_i; O_h = O \times C_i; T_h = T \times C_i$

For $v(g) = n, \Gamma = g$. For $v(u) = 2n, \Gamma = g$; for $v(u) = 2n + 1, \Gamma = u$.

The Approximations of Paul and van der Kelen

One defect of the Cotton–Kraihanzel field for species $M(CO)_5L$ is that it imposes two constraints on a problem with only one degree of freedom. A further defect is that it fails to allow for the detailed effects of ML bonding on interaction constants. An attempt has geen made by Paul and co-workers [345], and improved on by van der Kelen and colleagues, to remove these defects. The expressions used by van der Kelen are [346]

$$d(2 + y) = 3c$$

$$t(2 + y) = c(5 + y) \tag{IV.1}$$

where y is a parameter. These expressions may be derived from simple bonding arguments, in which the total π-bond order to carbon is assumed constant, as is the total charge on the metal. The y parameter is said to represent the π-acceptor capacity of L, and is fixed at unity for L = CO. This parameter can simply be eliminated from Eq. (IV.1), to give

$$c + d = t. \tag{IV.2}$$

Using (IV.2) as a constraint, the energy-factored force field may be solved to give

$$2d = K(B_2) - K(E)$$

$$t = k(2) - K(E)$$

$$c = t - d$$

$$k(1) = K(A_1)_1 + K(A_1)_2 + 2K(E) + 2K(B_2) - 4k(2)$$

$$k(1)[4k(2) - 2K(E) - K(B_2)] - [K(B_2) - K(E)]^2$$

$$- K(A_1)_1 . K(A_1)_2 = 0 \tag{IV.3}$$

Values of c, d and t for a range of Mn(I) complexes, for which exact (CO) parameters are available, are presented in Table IV.1. It is evident that Eq. (IV.2) is not invariably obeyed. In fact, agreement is sometimes very good and sometimes only fair, showing that it is dangerous to compare the van der Kelen parameters of different species in too much detail. However, it is evident that the parameters are an improvement on those of the CK approximation, and that if there is any value in approximate solutions to the energy factored force field, Eq. (IV.3) is at least as good as those of any other routinely applicable approximation.

TABLE 4.1. Comparison of c, d and t (Nm^{-1}) for Some Mn and Re Carbonyls (a).

Compound	c	d	t	$t/(c + d)$
$Mn(CO)_5Cl$	23·1	21·3	45·2	1·018
$Mn(CO)_5Br$	30·5	18·6	43·2	0·880
$Mn(CO)_5I$	28·6	18·1	41·8	0·895
$Mn(CO)_5CH_3$	31·0	24·3	47·4	0·839
$Re(CO)_5Cl$	25·0	28·1	58·6	1.104
$Re(CO)_5Br$	30·6	26·6	56·7	0·991
$Re(CO)_5I$	29·2	25·4	55·1	1·092
$Mn(CO)_5D$	25·7	30·4	48·5	0·865
$Re(CO)_5H(b)$	24·7	39·1	55·0	0·862
$Re(CO)_5H(c)$	28·2	32·8	58·3	0·956
$Re(CO)_5D$	28·1	33·0	58·2	0·953

(a) Data from (38, 299)
(b) Uncorrected for interaction with $v(M–H)$
(c) Corrected for interaction with $v(M–H)$

For *cis*-disubstituted species $M(CO)_4L_2$, the constraint used [374] is, as in [34, 35],

$$t = 2d \qquad (IV.4)$$

This may be incorporated into the "exact" equations

$$A_1: \begin{vmatrix} k(1) + c - K & 2d \\ 2d & k(2) + t - K \end{vmatrix} = 0 \qquad (IV.5)$$

$$B_1: \ k(1) - c = K$$

$$B_2: \ k(2) - t = K$$

to remove the one degree of freedom. The further assumption [34, 35] of equality between c and d is redundant; is not compatible even with the available data for unlabelled species (whatever relationship is assumed between d and t), and should therefore [38] be rejected.

Finally, it should be noted that the use of approximate expressions is not the best way of assigning spectra, for all species $cis - M(CO)_4L_2$, the $(A_1)_1$ band is at highest frequency and the B_1 band is lowest, but the order of $(A_1)_2$ and B_2 is sensitive to compound type and even [249] to solvent. The theory of intensities is very similar to that for species $mer - M(CO)_3L_3$ ($M(10)_3$, case iv); $(A_1)_1$ is weak in the infra-red, $(A_1)_2$ and B, of medium intensity, and B_2 strongest, while in the Raman spectrum, although all modes are formally active, B_2 is predicted to be weak.

References

1. See *e.g.* Braterman, P. S., *Structure and Bonding* **10,** 57 (1972).
2. Ransil, B. J., *Rev. Mod. Phys.,* **32,** 245 (1960).
3. Gilliland, W. L. and Blanchard, A. A., *J. Amer. Chem. Soc.* **48,** 410 (1926).
4. Pauling, L., "Nature of the Chemical Bond", Cornell University Press, 1939.
5. Jones, L. H., *In* "Advances in the Chemistry of the Coordination Compounds", (S. Kirschner, Ed.), Macmillan, New York, 1961, p. 398.
6. Cotton, F. A. *In* "Modern Coordination Chemistry", (Lewis, J., and Wilkins, R. G., Eds), Interscience, New York, 1960, p. 301.
7. Beach, N. A. and Gray, H. B., *J. Amer. Chem. Soc.* **90,** 5713 (1968).
8. Turner, D. W., Baker, C., Baker, A. D., and Brundle, C. R., "Molecular Photoelectron Spectroscopy", Interscience, London, 1970.
9. Siegbahn, K. *et al.,* "ESCA", Almqvist and Wiksells Bocktryckeri AB, Uppsala, 1967.
10. Farnell, L. F., Randall, E. W. and Rosenberg, E., *Chem. Comm.* p. 1078 (1971),
11. Gansow, O. A., Schexmayer, D. A., and Kimura, B. Y., *J. Amer. Chem. Soc.* **94,** 3406 (1972).
12. Born, M. and Oppenheimer, J. R., *Ann. Physik* **84,** 457 (1927).
13. Wilson, E. B., Jr., Decius, J. C., and Cross, P. R., "Molecular Vibrations", McGraw-Hill, New York, 1955.
14. Snyder, R. G., and Schachtschneider, J. H., *Spectrochim. Acta* **19,** 95 (1963); *idem, ibid,* p. 117.
15. Levenson, R. A., Gray, H. B., and Ceasar, G. P., *J. Amer. Chem. Soc.* **92,** 3653 (1970).
16. Richardson, J. W. Jr., "Organometallic Chemistry", (H. Zeiss, Ed.). Reinhold, New York, 1960.
17. Caulton, K. G., and Fenske, R. F., *Inorg. Chem.* **7,** 1273 (1968).
18. Brown, D. A., and Rawlinson, R. M., *J. Chem. Soc.* (*A*) p. 1530 (1969).
19. McLean, A. D. and Yoshimine, M., "Tables of Linear Molecule Wavefunctions", IBM, San Jose, 1967.
20. Cotton, F. A., Down, J. L., and Wilkinson, G., *J. Chem. Soc.* p. 833 (1959).
21. Huggins, D. K., and Kaesz, H. D., *J. Amer. Chem. Soc.* **86,** 2734 (1964).
22. Gilson, T. R., and Hendra, P. J., "Laser Raman Spectroscopy", Wiley, London, 1970.
23. Jones, L. H., McDowell, R. S., and Goldblatt, M., *Inorg. Chem.* **8,** 2349 (1969).
24. Bor, G., Johnson, B. F. G., Lewis, J., and Robinson, P. W., *J. Chem. Soc.* (*A*) p. 696 (1971).
25. Adams, D. M., "Metal-ligand and Related Vibrations", Edward Arnold, London, 1967.

258   REFERENCES

26. Jones, L. H., "Inorganic Vibrational Spectroscopy", Marcel Decker, New York, 1971, Vol. I, p. 157.
27. Abel, E. W., McLean, R. A. N., Tyfield, S. P., Braterman, P. S., and Walker, A. P., *J. Mol. Spectrosc.* **30**, 29 (1969).
28. Bor, G., *J. Organometal. Chem.* **10**, 343 (1967).
29. Bor, G., and Jung, G., *Inorg. Chim. Acta* **3**, 69 (1969).
30. Johnson, B. F. G., Lewis, J., Robinson, P. W., and Miller, J. R., *J. Chem. Soc. (A)* p. 1043 (1968).
31. Braterman, P. S., *J. Chem. Soc. (A)* p. 2907 (1968).
32. Braterman, P. S., *Chem. Comm.* p. 91 (1968).
33. Jones, L. H., *Inorg. Chem.* **8**, 1181 (1969).
34. Cotton, F. A., and Kraihanzel, C. S., *J. Amer. Chem. Soc.* **84**, 4432 (1962).
35. Kraihanzel, C. S., and Cotton, F. A., *Inorg. Chem.* **2**, 533 (1963).
36. Paul, I., Ph.D., Bristol University, 1968.
37. Dalton, J., Paul, I., Smith, J. G., and Stone, F. G. A., *J. Chem. Soc. (A)* p. 1195 (1968).
38. Braterman, P. S., Harill, R. W., and Kaesz, H. D., *J. Amer. Chem. Soc.* **89**, 2851 (1967).
39. Keeling, G., Kettle, S. F. A., and Paul, I., *J. Chem. Soc. (A)* p. 3143 (1971).
40. Braterman, P. S., Bau, R., and Kaesz, H. D., *Inorg. Chem.* **6**, 2097 (1967).
41. Darensbourg, D. J., *Inorg. Chim. Acta* **4**, 597 (1970).
42. Kettle, S. F. A., Paul, I., and Stamper, P. J., *Chem. Comm.* p. 1724 (1970).
43. Stolz, I. W., Dobson, G. R., and Sheline, R. K., *J. Amer. Chem. Soc.* **84**, 3589 (1962); *idem. ibid,* **85**, 1013, (1963).
44. Graham, M. A., Poliakoff, M., and Turner, J. J., *J. Chem. Soc. (A),* p. 2939 (1971).
45. Boylan, M. J., Braterman, P. S., and Fullarton, A., *J. Organometal. Chem.* **31**, C29 (1971).
46. Kettle, S. F. A., and Paul, I., *Inorg. Chim. Acta* **2**, 15 (1968).
47. Miller, J. R., *Inorg. Chim. Acta.* **2**, 421 (1968).
48. Pańkowski, M., and Bigorgne, M., *J. Organometal. Chem.* **19**, 393 (1969).
49. Nelson, N. J., Kime, N. E., and Shriver, D. F., *J. Amer. Chem. Soc.* **91**, 5713 (1969).
50. Marks, T. J., Kristoff, J. S., Alich, A. and Shriver, D. F., *J. Organometal. Chem.* **33**, C35 (1971).
51. Braterman, P. S., unpublished results.
52. Wei, C. H., *Inorg. Chem.* **8**, 2384 (1969).
53. Powell, H. M., and Ewens, R. V. G., *J. Chem. Soc.,* p. 286 (1939).
54. Neumann, M. A., Dahl, L. F., and King, R. B., cited by Penfold, B. R., *Perspectives in Structural Chem.* **2**, 109 (1968).
55. Corey, E. R., Dahl, L. F., and Beck, W., *J. Amer. Chem. Soc.* **85**, 1202 (1963).
56. Albano, V., Bellon, P. L., Chini, P., and Scatturin, V., *J. Organometal. Chem.* **16**, 461, (1969).
57. Bryan, R. F., and Manning, A. R., *Chem. Comm.* p. 1316 (1968); Ibers, J. A., *J. Organotmeal. Chem.* **14**, 423 (1968).
58. Wilson, F. C., and Shoemaker, D. P., *J. Chem. Phys.* **27**, 809 (1958).
59. Braterman, P. S., and Thompson, D. T., *J. Chem. Soc. (A),* p. 1454 (1968).
60. Dahl, L. F., and Wei, C. H., *Inorg. Chem.* **2**, 328 (1963).

61. Sly, W. G., *J. Amer. Chem. Soc.* **81**, 18 (1959).
62. Sumner, G. G., Klug, H. P., and Alexander, L. E., *Acta Cryst.* **17**, 732 (1964).
63. Wilkes, G. R., cited in Ref. 52.
64. Abel, E. W., Harrison, W., McLean, R. A. N., Marsh, W. C., and Trotter, J., *Chem. Comm.* p. 1513 (1970).
65. Kirtley, S. W., Olsen, J. P., and Bau, R., *J. Amer. Chem. Soc.* **95**, 4532 (1973).
66. Ref. 97, p. 1331.
67. Lewis, J., Manning, A. R., and Miller, J. R., *J. Chem. Soc. (A),* p. 845 (1966).
68. Manning, A. R., *J. Chem. Soc. (A),* p. 1135 (1968).
69. Kaesz, H. D., Smith, J. M., and Braterman, P. S., unpublished observations.
70. "Tables of Wavenumbers for the Calibration of Infra-Red Spectrometers", I.U.P.A.C., Butterworths, London, 1961.
71. Stiddard, M. H. B., private communication, 1962.
72. See e.g. Wender, I., Sternberg, H. W., and Orchin, M., *J. Amer. Chem. Soc.* **74**, 1216 (1952), Tucci, E. R., and Gwynn, B. H., *J. Amer. Chem. Soc.* **86**, 4838 (1964).
73. See *e.g.* Burlitch, J. M., and Ferrari, A., *Inorg. Chem.* **9**, 563 (1970).
74. Burlitch, J. M., *J. Amer. Chem. Soc.*, **91**, 4562 (1969).
75. Magee, T. A., Mathews, C. N., Wang, T. S., and Wotiz, J. H., *J. Amer. Chem. Soc.* **83**, 3200 (1961).
76. Jones, L. H., McDowell, R. S., and Goldblatt, M., *J. Chem. Phys.* **48**, 2663 (1968).
77. Black, J. B., and Braterman, P. S., *J. Organometal. Chem.* **39**, C (1972).
78. Poliakoff, M., and Turner, J. J., *J. Chem. Soc. (A),* p. 2403 (1971).
79. Hinchcliffe, A. J., Ogden, J. S., and Oswald, D. D., *J. C. S. Chem. Comm.,* p. 338 (1972).
80. Braterman, P. S., and Fullarton, A., *J. Organometal. Chem.* **31**, C27 (1971).
81. Graham, M. A., Perutz, R. N., Poliakoff, M., and Turner, J. J., *J. Organometal. Chem.* **34**, C34 (1972).
82. Graham, M. A., Ph.D. Cambridge, 1971, quoted in Ref. 44.
83. Ramsay, D. A., *J. Amer. Chem. Soc.* **74**, 72 (1952).
84. For a review see Saupe, A., *Angew, Chem. Internat. Edn.* **7**, 97 (1968).
85. Milne, D. W., Ph.D., Glasgow, 1970; Braterman, P. S., and Milne, D. W., to be published.
86. Buttery, H. J., Keeling, G., Kettle, S. F. A., Paul, I., and Stamper, F. J., *J. Chem. Soc. (A).* p. 471 (1970).
87. See e.g. Wood, E. A., "Crystals and Lights", Van Nostrand, Princeton, 1964.
88. Rigby, W., Whyman, R., and Wilding, K., *J. Phys. (E): Sci. Instrum.* **3**, 572 (1970).
89. Plyler, E. K., Baine, L. R., and Noack, M., *J. Res. Nat. Bureau Standards* **58**, 195 (1957).
90. Terzis, A., and Spiro, T. G., *Chem. Comm.,* p. 1160 (1970).
91. Jørgensen, C. K., "Absorption Spectra and Chemical Bonding in Complexes", Pergamon, London, 1962.
92. Edgar, K., Lewis, J., Manning, A. R., and Miller, J. R., *J. Chem. Soc. (A),* p. 1217 (1968).
93. Pince, R., and Poilblanc, R., *C. R. Acad. Sci. Paris, C.,* **272**, 83, (1971).
94. Smith, J. M., and Jones, L. H., *J. Mol. Spectrosc.* **20**, 248 (1966).
95. Dalton, J., Paul, I., and Stone, F. G. A., *J. Chem. Soc. (A),* p. 2744 (1969).

96. Lewis, J., Manning, A. R., Miller, J. R., Ware, M. J., and Nyman, F., *Nature* **207**, 142 (1965).
97. Cotton, F. A., and Wing, R. M., *Inorg. Chem.* **4**, 1328 (1965).
98. Vaska, L., *J. Amer. Chem. Soc.* **88**, 4100 (1966).
99. Ref. 76, pp. 2664, 2665.
100. Bouquet, G., and Bigorgne, M., *Spectrochim. Acta* **27A**, 139 (1971).
101. Manning, A. R., *J. Chem. Soc. (A)*, p. 1670 (1968).
102. Skinner, H. A., *Advan. Organometal. Chem.* **2**, 110 (1964).
103. Basolo, F., and Wojcicki, A., *J. Amer. Chem. Soc.* **83**, 520 (1961).
104. Greene, P. T., and Bryan, R. F., *J. Chem. Soc. (A)*, p. 1559 (1971).
105. For related structures see also La Placa, S. J., Hamilton, W. C., Ibers, J. A., and Davison, A., *Inorg. Chem.* **8**, 1928 (1969); Weber, H. P., and Bryan, R. F., *Acta Cryst.* **22**, 822 (1967); Einstein, F. W. B., Luth, H., and Trotter, J., *J. Chem. Soc. (A)*, p. 89 (1967); Bryan, R. F., *J. Chem. Soc. (A)*, p. 696 (1968).
106. Noack, K., *Helv. Chim. Acta* **45**, 1847 (1962).
107. Miller, J. R., *J. Chem. Soc. (A)*, p. 1855 (1971).
108. Bigorne, M., and Benlian, D., *Bull. Soc. Chim. France* p. 4100 (1967); Benlian, D., and Bigorgne, M., *ibid*, p. 4106.
109. Darensbourg, D. J., and Brown, T. L., *Inorg. Chem.* **7**, 959 (1968).
110. Kahn, O., *Ann. Chim. (France)*, **5**, 75 (1970).
111. El-Sayed, M. A., and Kaesz, H. D., *J. Mol. Spectrosc.* **9**, 310 (1962).
112. Ref. 134, pp. 3000–3001, 3014.
113. Cotton, F. A., *Inorg. Chem.* **3**, 702 (1964).
114. Graham, W. A. G., *Inorg. Chem.* **7**, 315 (1968).
115. Braterman, P. S., *4th Intl. Conf. Organometal Chem.* Bristol, August, 1969.
116. Burdett, J. K., *J. Chem. Soc. (A)*, p. 1195 (1971).
117. Bethke, G. W., and Wilson, M. K., *J. Chem. Phys.* **26**, 1118 (1957).
118. Armstrong, D. R., and Perkins, P. G., *J. Chem. Soc. (A)*, p. 1044 (1969).
119. Purcell, K. F., *Inorg. Chim. Acta* **3**, 540 (1969).
120. Fenske, R. F., XIII ICCC, Cracow, September 1970, *Pure App. Chem.* **27**, 61 (1971); Hall, M. B. and Fenske, R. F., *Inorg. Chem.* **11**, 1619 (1972).
121. Adams, D. M., and Newton, D. C., "Tables for Factor Group and Point Group Analysis", Beckmann—RIIC, London, 1970.
122. Adams, D. M., Hooper, M. A., and Squire, A., *J. Chem. Soc. (A)*, p. 71 (1971).
123. Buttery, H. J., Keeling, G., Kettle, S. F. A., Paul, I., and Stamper, P. J., *J. Chem. Soc. (A)*, p. 2077 (1969).
124. Lagemann, R. F., Nielsen, A. H., and Dickey, F. P. *Phys. Rev.* **72**, 284 (1947).
125. Hertzberg, G., and Rao, K. N., *J. Chem. Phys.* **17**, 1099 (1949).
126. Nakamoto, K., "Infrared Spectra of Inorganic and Coordination Compounds", 2nd Edn., Wiley-Interscience, New York, 1970.
127. Spielman, J. R., and Burg, A. B., *Inorg. Chem.* **2**, 1139 (1963).
128. Eischens, R. P., Pliskin, W. A., and Francis, S. A., *J. Chem. Phys.* **22**, 1786 (1964); *idem., J. Phys. Chem.* **60**, 194 (1965).
129. Seanor, D. A., and Amberg, C. H., *J. Chem. Phys.* **42**, 2967 (1965).
130. Haas, H., and Sheline, R. K., *J. Amer. Chem. Soc.* **88**, 3219 (1966).
131. Hieber, W., Peterhans, J., and Winter, E., *Chem. Ber.* **94**, 2572 (1961).
132. Kruck, Th., and Höfler, M., *Chem. Ber.* **97**, 2289 (1964).
133. Amster, R. L., Hannan, R. B., and Tobin, M. C., *Spectrochim. Acta* **19**, 1498 (1963).

134. Haas, H., and Sheline, R. K., *J. Chem. Phys.* **47**, 2996 (1967).
135. Abel, E. W., McLean, R. A. N., Norton, M. G., and Tyfield, S. P., *Chem. Comm.*, p. 900 (1968).
136. Beagley, B., Cruickshank, D. W. J., Pinder, P. M., Robiette, A. G., and Sheldrick, G. M., *Acta Cryst.* **B25**, 737 (1969).
137. Bor, G., *Inorg. Chim. Acta* **3**, 191 (1969).
138. Cataliotti, P., Foffani, A., and Marchetti, L., *Inorg. Chem.* **10**, 1594 (1971).
139. Bigorgne, M., *J. Organometal. Chem.* **24**, 211 (1970).
140. Calderazzo, F., and L'Eplattenier, P., *Inorg. Chem.* **6**, 1220 (1967).
141. Edgell, W. F., Huff, J., Thomas, J., Lehman, H., Angell, C., and Asato, G., *J. Amer. Chem. Soc.* **82**, 1254 (1960).
142. Stammreich, H., Sala, O., and Tavares, Y., *J. Chem. Phys.* **30**, 856 (1959).
143. Edgell, W. F., and Dunkle, M. P., *J. Phys. Chem.* **68**, 452 (1964).
144. Bouquet, G., and Bigorgne, M., *J. Mol. Structure* **1**, 211 (1967).
145. Stammreich, H., Kawai, K., Tavares, Y., Krumholz, P., Behmoiras, J., and Bril, S., *J. Chem. Phys.* **32**, 1482 (1960).
146. Edgell, W. F., and Lyford, J., IV, *J. Chem, Phys.* **52**, 4329 (1970).
147. Bor, G., *Chem. Comm.*, p. 641 (1969).
148. Flitcroft, N., Huggins, D. K., and Kaesz, H. D., *Inorg. Chem.* **3**, 1123 (1964).
149. Lindner, E., Behrens, H., and Birkle, S. *J. Organometal. Chem.* **15**, 165 (1968).
150. Dahl, L. F., and Rundle, R. E., *Acta Cryst.* **16**, 419 (1963).
151. Hardy, L. B., Ruff, J. K., and Dahl, L. F., *J. Amer. Chem. Soc.* **92**, 7312 (1970).
152. Griffith, W. P., and Wickham, A. J., *J. Chem. Soc. (A)*, p. 834 (1969).
153. Farmery, K. Kilner, M., Greatrex, R., and Greenwood, N. N., *Chem. Comm.*, p. 593 (1968).
154. Bor, G., *Sprectrochim. Acta* **19**, 2065 (1963).
155. Noack, K., *Spectrochim. Acta* **19**, 1925 (1963).
156. Noack, K., *Helv. Chim. Acta* **47**, 1555 (1964).
157. Bor, G., *Inorg. Chim. Acta* **3**, 196 (1969).
158. Rucci, G., Zangloterra, C., Lachi, M. P., and Camia, M., *Chem. Comm.*, p. 652 (1971).
159. Tate, D. P., Augl, J. M., Ritchey, W. M., Ross, B. L., and Grasselli, J. G., *J. Amer. Chem. Soc.* **86**, 3261 (1964).
160. Barraclough, C. G., and Lewis, J., *J. Chem. Soc.*, p. 4842, (1960).
161. Beck, W., and Lottes, K., *Chem. Ber.* **98**, 2657 (1965).
162. Behrens, H., Lindner, E., and Schindler, H., *Chem. Ber.* **99**, 2399 (1966).
163. Kruck, Th., and Lang, W., *Chem. Ber.* **98**, 3060 (1965).
164. Loutellier, A. and Bigorgne, M., *Bull. Soc. Chim. France*, p. 3186 (1965).
165. Bigorgne, M. *In* (S. Kirschner, Ed.), "Advances in the Chemistry of Coordination Compounds", Macmillan, New York, 1961, p. 199.
166. Bigorgne, M., *Bull, Soc. Chim. France,* p. 295 (1963).
167. Loutellier, A., and Bigorgne, M., *J. Chim. Physique* **67**, 99 (1970).
168. Bigorgne, M., *J. Organometal. Chem.* **1**, 101 (1963).
169. Kudo, K., Hidai, M., and Uchido, Y., *J. Organometal Chem.* **33**, 393 (1971).
170. Cotton, F. A., and Marks, T. J., *J. Amer. Chem. Soc.* **92**, 5114 (1970).
171. Faraone, F., Ferrara, C., and Rotondo, E., *J. Organometal. Chem.* **33**, 221 (1971).
172. Hieber, W., and Kummer, R., *Chem. Ber.* **100**, 148 (1967).
173. Vaska, L., and Peone, J., Jr., *Chem. Comm.*, p. 418 (1971).

174. Brown, C. K., Georgiou, D., and Wilkinson, G., *J. Chem. Soc.* (*A*), p. 3120 (1971).
175. Brown, C. K., and Wilkinson, G., *Chem. Comm.*, p.70 (1971).
176. Reed, C. A., and Roper, W. R., *Chem. Comm.*, p. 1556 (1971).
177. Irving, R. J., and Magnussen, E. A., *J. Chem. Soc.*, p. 1860 (1956); *idem, ibid*, p. 2283 (1958).
178. Denning, R. G., and Ware, M. J., *Spectrochim. Acta* **A24**, 1785 (1968).
179. Cardaci, G., and Murgia, S. M., *J. Organometal. Chem.* **25**, 483 (1970).
180. Koerner von Gustorf, E., Buchkremer, J., Pfaifer, Z., and Grevels, F.-W., *Angew. Chem. Intern. Ed.* **10**, 260 (1971).
181. Christian, D. F., and Roper, W. R., *Chem. Comm.*, p. 1271 (1971).
182. Hitch, R. R., Gondal, S. K., and Sears, C. T., *Chem. Comm.*, p. 777 (1971).
183. Moers, F. G., *Chem. Comm.*, p. 79, 1971.
184. Bercaw, J., Guastalla, G., and Halpern, J., *Chem. Comm.*, p. 1594, 1971.
185. Vallarino, L. M., *Inorg. Chem.* **4**, 161 (1965).
186. Deeming, A. J., and Shaw, B. L., *J. Chem. Soc.* (*A*), p. 376, (1971).
187. Bishop, J. J., and Davison, A., *Inorg. Chem.* **10**, 832 (1971).
188. Jetz, W., and Graham, W. A. G., *J. Amer. Chem. Soc.* **89**, 2773 (1967).
189. Brown, E. L., and Brown, D. B., *Chem. Comm.*, p. 67 (1971).
190. Shaw, B. L. and Smithies, A. C., *J. Chem. Soc.* (*A*), p. 2784 (1968).
191. Barbeau, C., *Canad, J. Chem.* **45**, 2 (1967).
192. Silverthorn, W. E., *Chem. Com.* **7**, 1310 (1971).
193. Cullen, W. R., Sams, W. R., and Thompson, J. A. J., *Inorg. Chem.* **10**, 843 (1971).
194. Green, M. L. H., Milchard, L. C., and Swannick, M. G., *J. Chem. Soc.* (*A*), p. 794 (1971).
195. Abel, E. W., and Moorhouse, S., *Inorg. Nucl. Chem. Letters* **7**, 905 (1971).
196. King, R. B., Treichel, P. M., and Stone, F. G. A., *J. Amer. Chem. Soc.* **83**, 3593 (1961).
197. Jetz, W., and Graham, W. A. G., *Inorg. Chem.* **10**, 4 (1971).
198. Oliver, A. J., and Graham, W. A. G., *Inorg. Chem.* **10**, 1 (1971).
199. King, R. B., and Efraty, A., *J. Organometal. Chem.* **27**, 409 (1971).
200. Rest, A. J., *Chem. Comm.*, p. 345 (1970).
201. Horrocks, W. D., jr., and Craig Taylor, R., *Inorg. Chem.* **2**, 723 (1963).
202. Thorsteinson, E. M., and Basolo, F., *J. Amer. Chem. Soc.* **88**, 3929 (1966).
203. Foà, M., and Cassar, L., *J. Organometallic Chem.* **30**, 123, (1971).
204. Olechowski, J. R., *J. Organometallic Chem.* **32**, 269 (1971).
205. Benlian, D., and Bigorgne, M., *Bull. Soc. Chim. France*, p. 1583 (1963).
206. Winkhaus, G., and Singer, H., *Chem. Ber.* **99**, 11 (1966).
207. Lawson, D. N., and Wilkinson, G., *J. Chem. Soc.* p. 1900 (1965).
208. Bonati, F., and Ugo, R., *J. Organometal. Chem.* **11**, 341 (1968).
209. Hieber, W., and Duchatsch, H., *Chem. Ber.* **98**, 2530 (1965).
210. Cotton, F. A., Liehr, A. D., and Wilkinson, G., *J. Inorg. Nucl. Chem.* **1**, 175 (1955).
211. Booth, B. L., Haszeldine, R. N., and Perkins, I., *J. Chem. Soc.* (*A*), p. 927 (1971).
212. Houk, L. W., and Dobson, G. R., *Inorg. Chem.* **5**, 2119 (1966).
213. Crossing, P. F., and Snow, M. R., *J. Chem. Soc.* (*A*), p. 610 (1971).
214. Behrens, H., Ruyter, E., and Lindner, E., *Z. Anorg. Allg. Chem.* **349**, 251 (1967).

215. Hales, L. A. W., and Irving, R. J., *Spectrochim, Acta* **23A**, 2981 (1967).
216. Reimann, R. H., and Singleton, E., *J. Organometal. Chem.* **32**, C44 (1971).
217. Poilblanc, R., and Bigorgne, M., *Bull. Soc. Chim. France,* p. 1301 (1962).
218. Pańkowski, M., and Bigorgne, M., *J. Organometal. Chem.* **30**, 227 (1971).
219. Johnson, B. F. G., Johnston, R. D., Lewis, J., and Williams, I. G., *J. Chem. Soc. (A),* p. 689 (1971).
220. Malatesta, L., Angoletta, M., and Conti, F., *J. Organometal. Chem.* **33**, 543 (1971).
221. Fischer, E. O., and Beck, H.-J., *Chem. Ber.* **104**, 3701 (1971).
222. Müller, J., and Fenderl, K., *Angew. Chem. Intern. Edn.* **10**, 418 (1971).
223. Höfler, M., and Schnitzler, M., *Chem. Ber.* **104**, 3117 (1971).
224. Strohmeier, W., and Hellmann, H., *Chem. Ber.* **98**, 1598 (1965).
225. Ruff, J. K., *Inorg. Chem.* **10**, 409 (1971).
226. Fischer, E. O., and Maasböl, A., *Chem. Ber.* **100**, 2445 (1967).
227. Green, M. L. H. and Wilkinson, G., *J. Chem. Soc.,* p. 4314 (1958).
228. Ugo, R., Cenini, S., and Bonati, F., *Inorg. Chim. Acta* **1**, 451, (1967).
229. Edmonson, R. C., and Newlands, M. J., *Chem. and Ind.* (London), p. 1888 (1966).
230. Sloan, T. E., and Wojcicki, A., *Inorg. Chem.* **7**, 1268 (1968).
231. King, R. B., and Bisnette, M. B., *J. Organometal. Chem.* **2**, 15 (1964).
232. Piper, T. S., and Wilkinson, G., *J. Inorg. Nucl. Chem.* **3**, 104 (1956).
233. Green, M. L. H., and Nagy, P. L. I., *J. Organometal. Chem.* **1**, 58 (1963).
234. King, R. B., Bisnette, M. B., and Fronzaglia, A., *J. Organometal. Chem.* **5**, 341 (1966).
235. Ariyaratne, J. K. P., and Green, M. L. H., *J. Chem. Soc.,* p. 1, (1964).
236. Davison, A., Green, M. L. H., and Wilkinson, G., *J. Chem. Soc.,* p. 3172 (1961).
237. Fischer, E. O., and Schneider, R. J. J., *Chem. Ber.* **103**, 3684 (1970).
238. Murray, J. G., *J. Amer. Chem. Soc.* **83**, 1287 (1961).
239. Calderazzo, F., and Bacciarelli, S., *Inorg. Chem.* **2**, 721 (1963).
240. Warren, J. D., and Clark, R. J., *Inorg. Chem.* **9**, 373 (1970).
241. Bailey, R. T., Lippincott, E. R., and Steele, D., *J. Amer. Chem. Soc.* **87**, 5346 (1965).
242. Murdoch, H. D., and Lucken, E. A. C., *Helv. Chim. Acta* **47**, 1517 (1964).
243. Casey, M., and Manning, A. R., *J. Chem. Soc. (A),* p. 256 (1971).
244. Cleland, A. J., Fieldhouse, S. A., Freeland, B. H., Mann, C. D. M., and O'Brien, R. J., *J. Chem. Soc. (A),* p. 736 (1971).
245. Cotton, F. A., and Parish, R. V., *J. Chem. Soc.,* p. 1440 (1960).
246. Hieber, W. Muschi, J., and Duchatsch, H., *Chem. Ber.* **98**, 3924 (1965).
247. Patmore, D. J., and Graham, W. A. G., *Inorg. Chem.* **7**, 771, (1968).
248. Whyman, R., *J. Organometal. Chem.* **29**, C36 (1971), and refs. therein.
249. Poilblanc, R., and Bigorgne, M., *J. Organometal. Chem.* **5**, 93 (1966).
250. Jones, C. E., and Coskran, K. J., *Inorg. Chem.* **10**, 55 (1971).
251. Barlow, C. G., Nixon, J. F., and Webster, M., *J. Chem. Soc. (A),* 2216 (1968).
252. Houk, L. W., and Dobson, G. R., *J. Chem. Soc. (A),* p. 317 (1966).
253. Dobson, G. R., and Houk, L. W., *Inorg. Chim. Acta* **1**, 287 (1967).
254. Kruck, Th., and Noack, M., *Chem. Ber.* **97**, 1693 (1964).
255. Ash, M. J., Brookes, A., Knox, S. A. R., and Stone, F. G. A., *J. Chem. Soc. (A),* p. 458 (1971).
256. Abel, E. W., Bennett, M. A., and Wilkinson, G., *J. Chem. Soc.,* p. 2323 (1959).

257. Bouquet, G., and Bigorgne, M., *Bull. Soc. Chim. France*, p. 433 (1962).
258. Osborne, A. G., and Stiddard, M. H. B., *J. Chem. Soc.*, p. 4715 (1962).
259. Abel, E. W., and Wilkinson, G., *J. Chem. Soc.*, p. 1501 (1959).
260. Fischer, R. D., *Chem. Ber.* **93**, 165 (1960).
261. Fritz, H. P., and Kreiter, C. G., *Chem. Ber.* **97**, 1398 (1964).
262. Öfele, K., and Dotzauer, E., *J. Organometal. Chem.* **30**, 211 (1971).
263. Parker, D. J., and Stiddard, M. H. B., *J. Chem. Soc.* (*A*), p. 2263 (1968).
264. Walker, P. J. C., and Mawby, R. J., *Inorg. Chem.* **10**, 404 (1971).
265. Jetz, W., and Graham, W. A. G., *Inorg. Chem.* **10**, 1159 (1971).
266. Schmid, G., and Nöth, H., *J. Organometal. Chem.* **7**, 129 (1967).
267. Patil, H. R. H., and Graham, W. A. G., *Inorg. Chem.* **5**, 1401 (1966).
268. Rest, A. J., *J. Organometal. Chem.* **25**, 530 (1970).
269. Seyferth, D., *J. Amer. Chem. Soc.* **82**, 1080 (1960).
270. Ruff, J. K., *Inorg. Chem.* **7**, 1499 (1968).
271. Edgell, W. F., Magee, C., and Gallup, G., *J. Amer. Chem. Soc.* **78**, 4185 (1956).
272. Bor, G., *Inorg. Chim. Acta* **1**, 82 (1967).
273. Kahn, O., Henrion, J., and Bouquet, G., *Bull. Soc. Chim. France,* p. 3547 (1967).
274. Watters, K. L., Butler, W. M., and Risen, W. M., Jr., *Inorg. Chem.* **10**, 1970 (1971).
275. Dalton, J., Paul, I., Smith, J. G., and Stone, F. G. A., *J. Chem. Soc.* (*A*), p. 1199, 1968, and references therein.
276. Davison, A., and Ellis, J. E., *J. Organometal, Chem.* **31**, 239 (1971).
277. Öfele, K., and Herberhold, M., *Angew. Chem. Intern. Edn.* **9**, 739 (1970).
278. Schlientz, W. J., Lavender, Y., Welcman, N., King, R. B., and Ruff, J. K., *J. Organometal. Chem.* **33**, 357 (1971).
279. Stolz, I. W., Dobson, G. R., and Sheline, R. K., *Inorg. Chem.* **2**, 323 (1963).
280. Stiddard, M. H. B., *J. Chem. Soc.* p. 4712 (1963).
281. Braterman, P. S., Wilson, V. A., and Joshi, K. K., *J. Chem. Soc.* (*A*), p. 191 (1971).
282. Hobday, M. D., and Smith, T. D., *J. Chem. Soc.* (*A*), p. 3424 (1971).
283. Braterman, P. S., Wilson, V. A., and Joshi, K. K., *J. Organometal. Chem.* **31**, 123 (1971).
284. Beck, W., Swoboda, P., Feldl, K., and Schuierer, E., *Chem. Ber.* **103**, 3591 (1970).
285. El-Sayed, M. A., and Kaesz, H. D., *Inorg. Chem.* **2**, 158 (1963).
286. Osborne, A. G., and Stone, F. G. A., *J. Chem. Soc.* (*A*), p. 1143 (1966).
287. Hayter, R. G., *J. Amer. Chem. Soc.* **86**, 823 (1964).
288. Angelici, R. J., *Inorg. Chem.* **3**, 1099 (1964).
289. L'Eplattenier, F., and Calderazzo, F., *Inorg. Chem.* **6**, 2092 (1967).
290. Calderazzo, F., *Inorg. Chem.* **4**, 223, (1965).
291. Calderazzo, F., *Inorg. Chem.* **5**, 429 (1966).
292. Bennett, M. A., and Clark, R. J. H., *J. Chem. Soc.* p. 5560 (1964).
293. Lindner, E., and Behrens, H., *Spectrochim. Acta* **A23**, 3025 (1967).
294. Wojcicki, A., and Farona, M. F., *J. Inorg. Nucl. Chem.* **26**, 2289 (1964).
295. Stolz, I. W., Dobson, G. R., and Sheline, R. K., *Inorg. Chem.* **2**, 1264 (1963).
296. Darensbourg, M. Y., and Darensbourg, D. J., *Inorg. Chem.* **9**, 32 (1970).
297. Fischer, E. C., Buthelt, W., and Müller, J., *Chem. Ber.* **104**, 986 (1971).
298. Brown, R. A., and Dobson, G. R., *J. Inorg. Nucl. Chem.* **33**, 892 (1971).

299. Kaesz, H. D., Bau, R., Hendrickson, D., and Smith, J. M., *J. Amer. Chem. Soc.* **89**, 2844 (1967).
300. Jetz, W., Simons, P. B., Thompson, J. A. J., and Graham, W. A. G., *Inorg. Chem.* **5**, 2217 (1966).
301. Cook, J., Green, M., and Stone, F. G. A., *J. Chem. Soc.* (*A*), p. 1973 (1968).
302. Farona, M. F., and Wojcicki, A., *Inorg. Chem.* **4**, 857 (1965).
303. Cariati, F., Romiti, P., and Valenti, V., *Gazz. Chim. Ital.* **98**, 615 (1968).
304. Manning, A. R., and Miller, J. R., *J. Chem. Soc.* (*A*). p. 3352 (1970).
305. Bor, G., *J. Organometal. Chem.* **11**, 195 (1968).
306. Bor, G., *Chem. Ber.* **96**, 2644 (1963).
307. Manning, A. R., *J. Chem. Soc.* (*A*), p. 1319 (1968).
308. Alich, A., Nelson, N. J., and Shriver, D. F., *Chem. Comm.* p. 254 (1971).
309. McArdle, P. A., and Manning, A. R., *Chem. Comm.,* p. 1020 (1968).
310. McArdle, P. A., and Manning, A. R., *J. Chem. Soc.* (*A*), p. 717 (1971).
311. Beveridge, A. D., and Clark, H. C., *J. Organometal. Chem.* **11**, 601 (1968), and references therein.
312. Marko, L., Bor, G., Klumpp, E., Marko, B., and Almoisy, G., *Chem. Ber.* **96**, 955 (1963).
313. Knight, J., and Mays, M. J., *Chem. Comm.,* p. 1006 (1970).
314. Poliakoff, M., and Turner, J. J., *Chem. Comm.,* p. 1008 (1970).
315. Evans, S., Green, J. C., Orchard, A. F., Saito, T., and Turner, D. W., *Chem. Phys. Letters* **4**, 361 (1969).
316. Hertzberg, G., "Spectra of Diatomic Molecules", Van Nostrand, Princeton, 1950.
317. Price, W. C., private communication.
318. Niewpoort, W. C., *Philips Res. Rept.* **20**, Suppl. 6 (1965).
319. Hillier, I. H., *J. Chem. Phys.* **52**, 1948 (1970).
320. Müller, J., Fenderl, K., and Mertschenk, B., *Chem. Ber.* **104**, 700 (1971).
321. Evans, S., Green, J. C., Green, M. L. H., Orchard, A. F., and Turner, D. W., *Disc. Faraday Soc.* **47**, 112 (1969).
322. Dewar, M. J. S., and Worley, S. D., *J. Chem. Phys.* **51**, 654 (1969).
323. Dewar, M. J. S., and Worley, S. D., *J. Chem. Phys.* **51**, 1672 (1969).
324. Barber, M., Connor, J. A., Hillier, I. H., and Saunders, V. R., *Chem. Comm.* p. 682 (1971); Clark, D. T., and Adams, D. B., *Chem. Comm.,* p. 740 (1971).
325. Hendrickson, D. N., Hollander, J. M., and Jolly, W. L., *Inorg. Chem.* **9**, 612 (1970).
326. Kramer, L. N., and Klein, M. P., *J. Chem. Phys.* **51**, 3618 (1969).
327. Barber, M., Connor, J. A., and Hillier, I. H., *Chem. Phys. Letters* **9**, 570 (1971).
328. Braterman, P. S., unpublished results.
329. Gray, H. B., and Beach, N., *J. Amer. Chem. Soc.* **85**, 2922 (1963).
330. Dartiguenave, M., Dartiguenave, Y., and Gray, H. B., *Bull. soc. Chim. France,* p. 4223 (1969).
331. Schreiner, A. F., and Brown, T. L., *J. Amer. Chem. Soc.* **90**, 3366 (1968).
332. Gray, H. B., Bernal, I., and Billig, E., *J. Amer. Chem. Soc.* **84**, 3404 (1962).
333. Braterman, P. S., and Walker, A. P., *Disc. Faraday Soc.* **47**, 112 (1969).
334. Brown, D. A., Chamber, W. J., Fitzpatrick, N. J., and Rawlinson, R. M., *J. Chem. Soc.* (*A*), p. 720 (1971).
335. Fenske, R. F., and DeKock, R. L., *Inorg. Chem.* **9**, 1053 (1970).
336. tom Dieck, H., and Renk, I. W., *Angew. Chem. Internat. Edn.* **9**, 793 (1970).

337. Gansow, O. A., Burke, A. R., and Lamar, G. N., *Chem. Comm.*, p. 456 (1971).
338. Braterman, P. S., Milne, D. W., Randall, E. W., and Rosenberg, E., J. C. S. Dalton, p. 1027 (1973).
339. Keiter, R. L., and Veʀkade, J. G., *Inorg. Chem.* **8**, 2115 (1969).
340. Mann, B. E., *Chem. Comm.*, p. 976 (1971).
341. Mann, B. E., *Chem. Comm.*, p. 1173 (1971).
342. Bramley, R., Figgis, B. N., and Nyholm, R. S., *Trans. Faraday Soc.* **58**, 1893 (1962).
343. (a) Rüber, H., Kündig, P., Moskovitz, M., and Ozin, G. A., *Nature Phys. Sci.* **235**, 98 (1972).
    (b) Kündig, P., Moskovitz, M., and Ozin, G. A., *J. Mol. Structure* **4**, 137 (1972).
344. Whitaker, A., and Jeffery, J. W., *Acta Cryst.* **23**, 977 (1967).
345. Dalton, J., Paul, I., and Stone, F. G. A., *J. Chem. Soc.* (A), p. 212 (1968).
346. Delbeke, F. T., Claeys, E. G., van der Kelen, G. P., and de Caluwe, R. M., *J. Organometal. Chem.* **23**, 497 (1970).
347. Breeze, P. A., and Turner, J. J., *J. Organometal. Chem.* **44**, C7 (1972).
348. Kettle, S. F. A., and Paul, I., *Advan. Organometal. Chem.* **10**, 199 (1972).
349. Darling, J. H., and Ogden, J. S., *J. C. S. Dalton* p. 2496 (1972).
350. Kettle, S. F. A., Paul, I., and Stamper, P. J., *J. C. S. Dalton* p. 2413 (1972).
351. Eliashev, M., and Wolkenstein, M., *J. Phys. (Moscow)* **9**, 101 (1945).
352. Mathieu, R., and Poilblanc, R., *Inorg. Chem.* **11**, 1858 (1972).
353. Fischer, E. O., and Herrmann, W. A., *Chem. Ber.* **105**, 186 (1972).
354. Labroue, D., and Poilblanc, R., *Inorg. Chem. Acta* **6**, 387 (1972).
353. Egdell, W. F., Yong, M. T., and Koizumi, N., *J. Amer. Chem. Soc.* **87**, 2563 (1965).
356. De Kock, C. W., and Van Leirsburg, D. A., *J. Amer. Chem. Soc.* **94**, 3235 (1972).
357. Blackmore, T., Cotton, J. D., Bruce, M. I., and Stone, F. G. A., *J. Chem. Soc.* (*A*) p. 2931 (1968).
358. Bruce, M. I., and Osterzewski, A. P. P., *Chem. Comm.*, p. 1124 (1972).
359. Bigorgne, M., and Pripathi, J. B. Pd., *J. Mol. Structure* **10**, 449 (1971).
360. Stewart, R. P., Okamoto, N., and Graham, W. A. G., *J. Organometal. Chem.* **42**, C32 (1972).
361. Clark, R. J., and Hoberman, P. I., *Inorg. Chem.* **4**, 1771 (1965).
362. Cavit, B. E., Grundy, K. R., and Roper, W. R., *J. C. S. Chem. Comm.*, p. 60 (1972).
363. Freni, M., Romiti, P., Valenti, V., and Fantucci, P., *J. Inorg. Nucl. Chem.* **34**, 1195 (1972).
364. Jeffery, J. and Mowby, R. J., *J. Organometal. Chem.* **40**, C43 (1972).
365. Pankowski, M., Demerseman, B., Bouquet, G., and Bigorgne, M., *J. Organometal. Chem.* **35**, 155 (1972).
366. Rehder, D., *J. Organometal. Chem.* **37**, 303 (1972).
367. van Bronswyk, W., and Clark, R. J. H., *Spectrochim. Acta* **28A**, 1429 (1972).
368. Bor, G., and Sbignadello, G., *J. C. S. Dalton* p. 440 (1974).
369. Cotton, F. A., Kruczynski, L., Shapiro, B. L. and Johnson, L. F., *J. Amer. Chem. Soc.* **94**, 6191 (1972).
370. Evans, J., Johnson, B. F. G., Lewis, J., Norton, J. R. and Cotton, F. A., *J. C. S. Chem. Comm.* p. 807 (1973).
371. Wei, C. H., Wilkes, G. R. and Dahl, L. F., *J. Amer. Chem. Soc.* **89**, 4792.

372. Johnson, B. F. G., Lewis, J. and Matheson, T. W., *J. C. S. Chem. Comm.* p. 441 (1974).
373. Lewis, J., Chem. Soc. Lecture, Glasgow University, 1974.
374. Delbeke, F. T., Claeys, E. G., de Caluwe, R. M. and van der Kelen, G. P., *J. Organometal. Chem.* **23,** 505 (1970).
375. Knox, S. A. R., Hoxmeier, R. J. and Kaesz, H. D., *Inorg. Chem.* **10,** 2636 (1971).
376. Mays, M. J. and Stefanini, F. P., *J. Chem. Soc. (A)* p. 2747 (1971).
377. Jetz, W. and Graham, W. A. G., *J. Amer. Chem. Soc.* **91,** 3375 (1969).
378. Hartmann, F. A. and Wojcicki, A., *J. Amer. Chem. Soc.* **88,** 844 (1966).

Author Index

The numbers in brackets refer to the Reference Section where the references are listed in full.

A

Abel, E. W., 34 (27), 87 (64), 174 (27), 180 (27), 181 (27), 182 (27), 184 (27, 135), 193 (27), 203 (195), 214 (256, 259), 237 (27), 239 (27)
Adams, D. B., 233 (324)
Adams, D. M., 32 (25), 175 (121), 176 (122), 192 (122), 194 (122)
Albano, V., 78 (56), 79 (56)
Alexander, L. E., 86 (62), 196 (62)
Alich, A., 77 (50), 225 (308), 241 (50)
Almoisy, G., 226 (312)
Amberg, C. H., 179 (129)
Amster, R. L., 180 (133), 181 (133)
Angelici, R. J. 219 (288)
Angell, C., 186 (141)
Angoletta, M., 208 (220)
Ariyaratne, J. K. P., 210 (235), 220 (235)
Armstrong, D. R., 172 (118)
Asato, G., 186 (141)
Ash, M. J., 214 (255)
Augl, J. M., 199 (159)

B

Bacciarelli, S., 210 (239)
Bailey, R. T., 212 (241)
Baine, L. R., 136 (89)
Baker, A. D., 2 (8), 230 (8), 231 (8), 232 (8)
Baker, C., 2 (8), 230 (8), 231 (8), 232 (8)
Barbeau, C., 203 (191)
Barber, M., 233 (324), 234 (327)

Barlow, C. G., 213 (251)
Barraclough, C. G., 199 (160)
Basolo, F., 167 (103), 206 (202), 211 (202)
Bau, R., 40 (40), 70 (40), 88 (65), 159 (299), 169 (40), 223 (299), 226 (40), 255 (299)
Beach, N. A., 2 (7), 237 (7), 238 (7), 239 (7, 329)
Beagley, B., 185 (136)
Beck, H. J., 208 (221)
Beck, W., 78 (55), 199 (161), 206 (161), 207 (161), 211 (161), 212 (161), 219 (284)
Behmoiras, J., 190 (145), 191 (145)
Behrens, H., 192 (149), 199 (162), 206 (162), 208 (214), 209 (214), 212 (162), 214 (214), 221 (293)
Bellon, P. L., 78 (56), 79 (56)
Benlian, D., 169 (108), 172 (108), 207 (205), 212 (205)
Bennett, M. A., 214 (256), 221 (292)
Bercaw, J., 202 (184), 207 (184)
Bernal, I., 239 (332)
Bethke, G. W., 172 (117), 178 (117)
Beveridge, A. D., 225 (311)
Bigorgne, M., 72 (48), 166 (100), 169 (108), 172 (108), 186 (139), 187 (139), 188 (139), 189 (100, 139, 144), 190 (100), 191 (100), 199 (164, 165, 166, 167, 168), 202 (359), 206 (164, 165, 167), 207 (168, 205), 208 (217, 218), 212 (164, 165, 167, 168, 205), 213 (139, 249), 214 (166, 257), 217 (139), 218 (249, 257), 219 (168, 257), 222 (168, 217), 223 (217), 256 (249)

269

W

Walker, P. J. C., 215 (264)
Wang, T. S., 119 (75)
Ware, M. J., 162 (96), 201 (178)
Warren, J. D., 212 (240)
Watters, K. L., 217 (274)
Weber, H. P., 167 (105)
Webster, M., 213 (251)
Wei, C. H., 78 (52), 86 (60), 243 (371)
Welchman, N., 218 (278), 221 (278)
Wender, I., 115 (72)
Whitaker A., (344)
Whyman, R., 135 (88), 213 (248)
Wickham, A. J., 195 (152)
Wilding, K., 135 (88)
Wilkes, G. R., 87 (63), 243 (371)
Wilkinson, G., 21 (20), 201 (174, 175), 202 (174), 203 (174, 175), 207 (207, 210), 209 (227, 232), 210 (236), 214 (256, 259), 216 (232)
Williams, I. G., 208 (219)
Wilson, E. B., Jr. 6 (13), 9 (13), 13 (13), 14 (13), 20 (13), 21 (13), 22 (13), 25 (13), 247 (13)

Wilson, F. C., 83 (58), 85 (58)
Wilson, M. K., 172 (117), 178 (117)
Wilson, Y. A., 219 (281, 283)
Wing, R. M., 93 (66, 97), 162 (97), 193 (97)
Winkhaus, G., 207 (206)
Winter, E., 180 (131), 217 (131), 212 (131)
Wojcicki, A., 167 (103), 209 (230), 215 (230), 221 (294), 224 (302, 378)
Wolkenstein, M., 169 (351)
Wood, E. A., 134 (87)
Worley, S. D., 233 (322, 323)
Wotiz, J. H., 119 (75)

Y

Yong, M. T., 190 (355)
Yoshimine, M., 21 (19), 168 (19)

Z

Zangloterra, C., 199 (158)

Subject Index

Compound Index